10 Technologies Every Executive Needs to Know

Dermot McCormack and Fergal O'Byrne

If you are a C-Level executive interested in submitting a manuscript to the Aspatore editorial board, please email jason@aspatore.com. Include your book idea, your biography, and any additional pertinent information.

Published by Aspatore, Inc.

For corrections, company/title updates, comments or any other inquiries, please e-mail info@aspatore.com.

First Printing, 2004

10 9 8 7 6 5 4 3 2 1

Copyright © 2004 by Aspatore, Inc. All rights reserved. Printed in the United States of America. No part of this publication may be reproduced or distributed in any form or by any means, or stored in a database or retrieval system, except as permitted under Sections 107 or 108 of the United States Copyright Act, without prior written permission of the publisher.

ISBN 1-58762-368-4 Library of Congress Control Number: 2004101126

Material in this book is for educational purposes only. This book is sold with the understanding that neither any of the authors or the publisher is engaged in rendering medical, legal, accounting, investment, or any other professional service. For legal advice, please consult your personal lawyer.

This book is printed on acid free paper.

The views expressed by the individuals in this book do not necessarily reflect the views shared by the companies they are employed by (or the companies mentioned in this book).

If you are interested in purchasing bulk copies for your team/company with your company logo and adding content to the beginning of the book, or licensing the content in this book (for publications, web sites, educational materials), or for sponsorship, promotions or advertising opportunities, please email store@aspatore.com or call toll free 1-866-Aspatore.

About ASPATORE BOOKS –
Publishers of C-Level Business Intelligence

www.Aspatore.com

Aspatore Books is the largest and most exclusive publisher of C-Level executives (CEO, CFO, CTO, CMO, Partner) from the world's most respected companies. Aspatore annually publishes a select group of C-Level executives from the Global 1,000, top 250 professional services firms, law firms (Partners & Chairs), and other leading companies of all sizes. C-Level Business Intelligence ™, as conceptualized and developed by Aspatore Books, provides professionals of all levels with proven business intelligence from industry insiders – direct and unfiltered insight from those who know it best – as opposed to third-party accounts offered by unknown authors and analysts. Aspatore Books is committed to publishing a highly innovative line of business books, and redefining such resources as indispensable tools for all professionals.

This Book is dedicated to Marco, Doran and Jo Ann

— **Dermot A. McCormack**

I dedicate this book to my beautiful Danish wife, Maiken, without whom I would never have known where the name for Bluetooth came from! Also, to our two cats Orbix and Polaris for insisting on walking across my keyboard and thus adding greatly to the content of this book.

— **Fergal O'Byrne**

Table of Contents

1. WEB SERVICES..3
2. NANOTECHNOLOGY 15
3. SECURITY 25
4. GRID COMPUTING..37
5. LINUX...49
6. WIRELESS TECHNOLOGIES 63
7. XML 75
8. CUSTOMER RELATIONSHIP MANAGEMENT (CRM)..87
9. J2EE 101
10. MICROSOFT.NET...111
APPENDIX: GLOSSARY OF TERMS

Introduction

With every day that passes in this constantly changing business landscape, technology continues to become more and more entwined in everything we do. From the Internet to wireless networks to data warehousing to security, today's executive is expected to go beyond the buzzwords and really understand how these technologies can affect his or her business and how to maximize technology investment; the alternative is to be held hostage by these technologies or even worse, by your own IT department!

In this book, we try to act as your technology ambassador, simply and clearly explaining the core concepts behind and the possible ramifications of 10 of these technologies which I believe will undoubtedly have lasting effects on your organization in the coming years.

Too many books on technology tend to start out with good intentions, promising a clear and concise explanation and roadmap but all too often end up requiring the reader to have a degree in astrophysics to get past the first chapter. We've tried to keep this book simple and informative with descriptions in plain English and plenty of diagrams and we even have a thorough glossary of terms for each section. And in the likely event that you need a quick cheat sheet before that next meeting, we added a "Top Ten Things You Need to Know About." cheat sheet and a simple "Bottom Line" paragraph on each topic.

We hope this book provides you with at least a top line understanding of all these new technologies and their associated jargon, and you view this book as a concise travel guide to that strange country "Technology" that you will now be visiting more and more.

Who knows, maybe you'll even become a citizen one day!

Enjoy!

Dermot A. McCormack

Fergal O'Byrne

1

Web Services

"The difference between 'involvement' and 'commitment' is like an eggs-and-ham breakfast: the chicken was 'involved' - the pig was 'committed.'"

- Unknown

The fact is, just as the Internet did, Web services are likely to become a very important part of the business landscape over the next few years. So what does this mean for you and your business and, while we are at it, what are Web Services anyway? Read on.

Web Services are basically a set of tools and protocols, which enable software applications to communicate, pass data and issue commands to each other over the Internet or any other network. Web Services are a kind of plumbing that connects different programs together using a network and can be used both inside the organization and to integrate with other organizations.

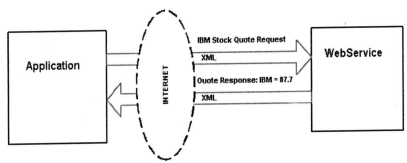

Figure 1.1: Simple Web Services example

Let's imagine you have a piece of software you want others to use. That piece of software may initiate or accept business transactions, it may provide information or it may manage one of your many business processes. You may want to let another piece of software communicate with your software. The communication may be internal, from another application inside the company, it may be external, from one of your suppliers or customers say. Web Services provide a way to connect these applications together and allow them to pass the required pieces of data in a simple standardized way while being secure and reliable (at least that's the hope!). Web services also define how another company may find your piece of software (or service) and offer instructions on how to connect to it.

It is important to remember that Web services are still a work in progress and many of the standards are new and not widely tested. There are also unresolved issues surrounding security and reliability. Having said all that it is likely that many of these issues will be ironed out and it is also likely that there are several smaller Web Service initiatives already underway in your organization right now.

It is important to start planning for Web services now and understand what current vendors plans are for Web Services support and integration.

Detailed Summary:

Let's take a brief history lesson in Enterprise computing applications. From as early as the 1960s through the late 1970s, Enterprise computing applications were simple in design and functionality, developed largely in part to end repetitive tasks. There was no real thought whatsoever given to the integration of corporate data. The entire objective was to replicate manual procedures on the computer. By the 1980s, several corporations were beginning to understand the value and necessity for application integration. As ERP (Enterprise Resource Planning) applications became much more prevalent in the 1990s, there was a need for corporations to be able to leverage already existing applications and data within the ERP system. Basically data and systems grew up as islands and eventually there was a need to connect them. Thus System Integration was born (much to the chagrin of people who've had to pay for it over the years!). Web Services can be viewed as the latest incarnation of system integration leveraging the networking phenomenon we know as the Internet.

Let's start with some definitions from some of the larger players in the Web services arena:

Web Services

"A Web service is an interface that describes a collection of operations that are network accessible through standardized XML messaging. Self-contained, modular applications that can be described, published, located and invoked over a network (generally the Internet)."

[IBM]

"A Web Service is programmable application logic accessible using standard Internet protocols. Web services combine the best aspects of component-based development and the Web."

[MICROSOFT]

"Web services are self-describing applications that can discover and engage other web applications to complete complex tasks over the Internet."

[SUN]

So Web Services allow software applications to both talk to each other and pass data and commands back and forth both inside and outside the company.

This all sounds great, I hear you say, but haven't we done all this before and haven't we spent a lot of money on making all these pieces of software talk to each other? What about all those technologies like MQSeries ™, Vitria ™, Middleware and Java that we have, don't they do this?

The short answer is YES and YES we have done all this with all these before with these technologies but in a hundred different ways and in no particular adherence to any set of standards which makes every integration a standalone project unto itself. Most systems today are integrated based on technologies such as message queuing and buzzwords like Enterprise Architecture Integration (EAI) and Business Process Integration (BPI), but many of the standards involved are vendor specific and not necessarily conducive to true interoperability.

One of the core ideas behind Web Services it that it reduces the complexity of business integration and offers companies the ability to use a core set of standards and best-of-breed technologies. This means you can leverage the many investments in Internet technologies and training you may have made over the last few years and allows IT managers to spend more time on the underlying infrastructure that makes your business more efficient and less time on the actual "plumbing."

Let's look at a simple example: You run a progressive Auto dealership and receive an order for a new vehicle via your website. The first thing you do is run a credit check and submit a loan application via a Web service from your financial services provider. Then it is likely you would check availability from your local inventory database and, assuming you didn't have that particular model in stock, you would query the manufacturer's central database, again via Web service. Assuming success, you would connect to the Insurance Company's Web service and arrange coverage for the automobile via yet another Web service. (See Figure 2)

Get the picture? In other words you are using Web services to streamline communication, workflow and business process automation.

Web Services 7

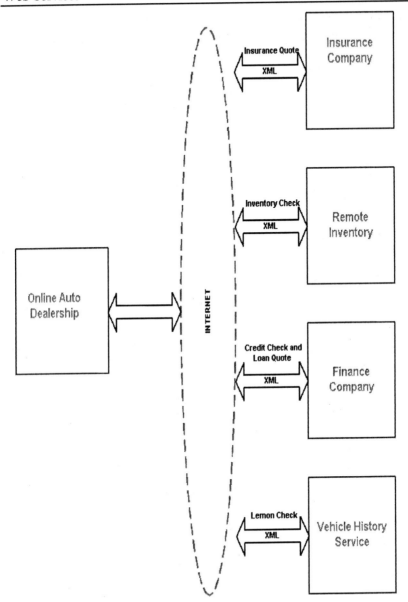

Figure 1.2: Online Auto Dealership Example

Web Services is, at its simplest, an XML transfer over HTTP protocol. This is the message and the transport mechanism. As we get deeper in the specification we encounter the concept on an "envelope" or SOAP (Simple Object Access Protocol), which basically adds levels of protection and accountability to the transaction. We also see directory services and listing protocols so companies can, in theory, automatically search directories for services provided by other companies and automatically negotiate for and use these services. Protocols like eBXML (electronic business XML) essentially round out the picture by creating a secure, robust and "corporate" standard for organizations to connect to each other via Web Services.

From the www.ebXML.org:

"ebXML (Electronic Business using eXtensible Markup Language), is a modular suite of specifications that enables enterprises of any size and in any geographical location to conduct business over the Internet. Using ebXML, companies now have a standard method to exchange business messages, conduct trading relationships, communicate data in common terms and define and register business processes."

The Web services momentum we are seeing is largely based on the fact that the advantages on the Web as a platform, i.e., its simplicity and ubiquity, apply not only to information but also to services. Having said all that, the vision for Web services is truly admirable but we have a ways to go before we realize the true potential of this vision. There is much work to be done in the area of standards and more important the acceptance of these standards by the development community. One thing the web service movement has to be careful of is large vendors like IBM and Microsoft splintering off and developing their own set of standards. This didn't happen with the Web but certain vendors may see this as an opportunity to "own" part of the growing Web Services standard having missed out on the ability to own and control the initial Web standards.

How will Web Services Affect Businesses?

One thing to keep in mind about web services is that, no matter whom you listen to, the sector *will* come of age and *will* ultimately change the face of businesses' integration and how information and data is exchanged.

Web Services are the next logical wave of the Internet revolution. Applications will be linked together the way documents were in the first phase of the Web and, as with the first phase, we are likely to see ramifications far beyond what we expected.

Of course the rise of the web services platform with the promise of allowing near seamless integration of almost any application on any platform raises some serious questions for the Vendor community.

If everyone speaks the same language, who needs a translator?

Integration, in theory, becomes easier allowing companies to adopt best of breed strategies instead of being locked into entire suites.

An important thing to remember is that we do not have to reference specific operating systems, programming languages, hardware, databases, CRM systems or anything like that to define Web services integration. This gives organizations the flexibility to put a service on the platform that makes the most business sense and delivers the quality of service customers require. This may lead to a reduction in dependency on any one single vendor which in turn lends the IT manger more bargaining power.

One thing we do know, if Web Services are widely adopted, integration will become easier. So who are the winners and losers here? The customer gets better integrated and more flexible systems, the niche vendor sells more software and the system integrators do more best of breed integrations, clearly all winners. The losers may well be the large Suite vendors who will no longer be able to sell the customer on an entire solution with certain weak components, which may explain why the bigger players are so hunger to "own" a piece of the web services standards.

There's no doubt Web Service will play an important role in the future of software development and computing in general but the question is "is it really ready for prime time" and is there an immediate provable ROI in implementing these services in the short term. Many business managers and IT professions are still a little gun-shy when it comes to new technology investments and the boom and bust of the nineties, but that doesn't mean you should ignore web services and their possibilities.

Web services have a larger relevance for applications both inside and outside the company (or inside and outside the firewall as IT managers are fond of saying). Being able to move one's data efficiently around the organization is becoming more and more important in today's agile enterprise and the widespread adoption of Web Service is likely to play a key role in "moving data" in the organization. The whole idea of Web services is to get dissimilar systems, platforms, and applications to talk to each other and basically exchange not just data but also commands.

Remember Web services are not the be all and end all. Web services can be viewed as the middle men. Without robust, relevant and functional applications behind them they become quite irrelevant. Companies should view web services as a complement to what they already have; a way to get more out of the stuff you've already invested in, NOT as a replacement!

What to Watch for in the Future:

One must remember that the world of Web Services is a relatively new one and many of the pieces that purport to make up the standard are still in their developmental stages. All in all, Web services have a ways to go before it becomes a hard baked, viable standard for corporations to use.

The current push for standards is taking a top down, bottom up approach.

On the top you have the large corporations huddling together to come up with complex protocols and standards and on the bottom you have the grass roots developer community moving the ball forward with XML and SOAP applications popping up every day.

A lot of the progress we have seen to date and will see in the foreseeable future around Web Services will happen "behind the firewall," i.e., inside the organization as companies move to pry customer data from a backend legacy database to front end web applications, and attempt to integrate more and more disparate systems. Over time, Web Services implementations will move outside the firewall as the standards evolve and more companies feel comfortable exchanging their data and transaction via this method.

It is also highly likely that Web Services will play a large role in the move towards GRID/Distributed computing and OnDemand computing (See Chapter 4) as the internet starts to actually turn into the computer. As Sun Microsystems pointed out in one of their ads in the nineties, "The Network is The Computer." Maybe they were right.

Web Services

Most Frequently Asked Questions:

What are Web Services?

Web Services are basically a set of tools and protocols, which enable software applications to communicate, pass data and issue commands to each other over the Internet or any other network. Web Services are a kind of plumbing that connects different programs together using a network and can be used both inside the organization and to integrate with other organizations.

What's the difference between a Web service and a website?

Unlike websites, which are pages designed to be viewed in a browser by a person, a Web service is designed to be accessed directly by another service or software application.

Why do I need Web Services?

There's a pretty good chance your business is dependant on its IT systems and the data steams those systems produce. Web services are likely to become the standard for connecting and integrating these systems and the conduit for converging all these data streams in one place.

When are Web Services likely to be commonplace in the enterprise?

We are probably looking at 2 to 4 years before we start seeing real momentum.

What should I do in the meantime?

This best bet, in the meantime, before Web services become rock solid for intra company communication is to sponsor internal web services based applications. This will give your IT group a leg up on Web services technology and generate real ROI and efficiencies in the short term if done right.

Don't we have this already with the millions we have spent on IT?

You probably have several different versions of this already and that's partly the problem. If you don't envision more system integration in your company or have a completely homogeneous set of computing platforms then you are probably ok. If not then you might need to look at Web services. The nice thing about getting started is that you can reuse a lot of your existing technologies.

Estimated Timeline:

As with most new technologies and paradigms, there is no one date where they finally arrive and are completely adopted. There are many applications out there today that are essentially Web Services or use Web Service principles.

Buzz Words and Definitions:

Web services are built using a sea of acronyms that you will at least have to have a general understanding of, so here are a few of the more important protocols that make up Web Services.

XML (*Extensible Markup Language*)- This is a customizable way of creating tags to enable the definition, transmission, validation, and interpretation of data between applications and between organizations. It is essentially the format the web services document or request is in.

SOAP (*Simple Object Access Protocol*) - SOAP is a messaging protocol used to encode XML-based messages over the Internet. It adds and defines security and transactional information associated with the Web Services document or request.

UDDI (*Universal Description, Discovery and Integration*)- A Web-based distributed directory that enables businesses to list themselves on the Internet and discover each other, similar to a traditional phone book's yellow and white pages. Basically, a directory where companies can find the service I'm looking for and list the services they provide in a Web Services format.

WSDL (*Web Services Description Language*) - An XML-formatted language used to describe a Web service's capabilities as collections of communication endpoints capable of exchanging messages. WSDL is an integral part of

Web Services

UDDI, an XML-based worldwide business registry. WSDL is the language that UDDI uses. Microsoft and IBM developed WSDL jointly. In plain English, when I do find the service I'm looking for, this explains what it is and how to connect to it.

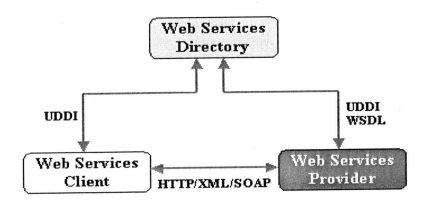

FIGURE 1.3: Web Services Request Flow Example

ebXML (electronic business Extensible Markup Language)- A modular suite of specifications for standardizing XML globally in order to facilitate trade between organizations regardless of size. The specification gives businesses a standard method to exchange XML-based business messages, conduct trading relationships, communicate data in common terms and define and register business processes.

 The Bottom Line:

You will deploy based on what system and tools you already have invested in and where the experience level of your tech team lays. And ultimately, if this thing is done right, it shouldn't really matter what system you choose as it will all speak the same language anyhow (hopefully).

Top Ten Cheat Sheet: Web Services

- Think plumbing between computer programs to pass data and commands

- Web services are the next logical step in integrating the Internet into the Enterprise

- The standards are not really baked yet and many things still need to be agreed on before Web Services can really take off

- XML is at the heart of Web services

- Chances are your company is already using Web Services in some format. If not, they should start

- If implemented correctly, Web Services should greatly decrease the complexity of system integration for organizations

- Web Services will be THE protocol for data exchange over the Internet

- Sun Microsystems's J2EE and Microsoft's .NET are currently the two main development platform choices for Web Services

- Most work with Web Services today is done "intra organization".

- The next growth phase will be in the realm of company-to-company communication where different organizations will purchase, negotiate, sell, find and provide services and share data with each other using Web services.

2

Nanotechnology

"There are only two tragedies in life: one is not getting what one wants, and the other is getting it."

- Oscar Wilde (1854-1900)

A nanometer is one billionth of a meter – about the thickness of six atoms placed side to side. A single strand of human hair is approximately 50,000 to 100,000 nanometers! Nanotechnology is all about using "machines" and building things from 1 to a few 100 nanometers. This is extremely small-scale stuff. I would like you to cast your mind back to the halcyon days of other small-scale technologies. First, there was the humble transistor – fabricated from discrete parts and sitting on a board along with inductors and capacitors. Then came the printed circuit board, reducing the size requirements considerably.

With the advent of photolithography (A form of lithography in which light-sensitive plates or stones are exposed to a photographic image, usually by means of a halftone screen), these circuits were reduced down to the micron level (one millionth of a meter). It is worth remembering Moore's Law - first coined by George Moore in 1965, the future chairman and CEO of Intel - when he stated that the number of transistors that could be fitted onto an integrated circuit had doubled every year since 1961. Nowadays experts agree this figure is nearer to 18 months. This cannot go on indefinitely, of course, but many experts feel we are reaching the boundaries of what can be achieved with current lithographic technology. This is where Nanotechnology enters the fray.

Let's start with some basics. Everything, including you and I, are made from atoms. All manufactured products are merely very complex arrangements of atoms. If we rearrange the atoms in coal we can make diamond. If we rearrange the atoms in sand we can, eventually, make computer chips. By arranging atoms of soil, water and air we can make potato chips.

If we are to be honest with ourselves, most of the products that we make today are still very unsophisticated, at a molecular level. We do not, yet, have the ability to move things around at an atomic level. Even the current processes we use to make the integrated circuits used in today's computers and processors rely on complex statistical formulae for moving very large numbers of atoms around in certain temperature environments. We are atom herders and not atomic builders. Nanotechnology is all about having the ability to control individual (or small groups) of atoms and to get them to perform tasks and do what we want them to do. One excellent analogy of what we are doing today is that we are trying to play pick-up-sticks with boxing gloves on – it is not pretty or easy – but it can be achieved. Using this technique we could, maybe, move piles of sticks from one place to another. Using nanotechnology we should be able to make those same sticks into a model of a galleon ship.

Detailed Summary:

Nanotechnology will enable us to snap together the fundamental building blocks of nature in an easy and inexpensive manner. The computer revolution brought about by the ability to work at the micron level will be nothing compared to what can be achieved if we can work at the nanometer level. This is truly a major leap in technology. If you think about a nanometer as being approximately six atoms laying side by side you can appreciate the scales we are discussing. The actual use of the word nanotechnology does include dimensions up to about 1,000 nanometers – so there is a lot of scope to work with here.

Some people working in the lithography field would claim to be working in the nanometer space – but nanotechnology is an entire philosophy of working on very, very small scales. Lithography will always flounder if we try to go smaller and smaller. Current technology in this field does not let us "tell" an atom of arsenic where it should be within a germanium lattice – it is all worked on probabilities and statistical analysis – noting that a certain number of dopant atoms (arsenic in this case) should be present in a given sample of germanium. Pundits agree that conventional lithography is starting to reach the limits of how much further it can be reduced in size.

Nanotechnology will enable us to build computer systems using building blocks fabricated from molecular amounts of logic elements. Imagine the ability to fabricate transistors and their associated tracking and conductors

Nanotechnology

with elements that are literally atoms wide! The possibilities are boundless. Nanotechnology is touted as the technology that will make this happen.

So why are so many people putting so much time, money and resources into this area (billions in research dollars have already been invested by governments and firms alike) – it sounds too good to be true. The fact is, if achieved, it will be too good to be true. It will provide manufacturers with an extremely efficient way to take raw materials (atoms, in effect) and combine them to make complex machines. The efficiencies to be gained are massive. In a few decades, this emerging manufacturing technology will let us arrange atoms and molecules in most of the ways permitted by physical law. It will let us make supercomputers that fit on the head of a pin and fleets of nano-robots able to eliminate clogged arteries in the human body.

Besides computers billions of times more powerful than those used today, and new medical capabilities that will heal and cure, this new and very precise way of fabricating products will also eliminate the pollution from current manufacturing methods. Molecular manufacturing will be pollutant free. In theory there should be no waste.

The origins of nanotechnology start with the pioneer Eric Drexler, who first published his vision in the book "The Engines of Creation" in 1986. The response from the scientific community could be described as skeptical, at best. Some commentators were overtly hostile to the ideas promulgated by Drexler. It seemed too good to be true, and many scientists pronounced the whole thing impossible. The main gripe of the detractors was that it would be physically impossible to manipulate single atoms and to build machines at the molecular level.

Drexler must feel somewhat vindicated by the fact that in 2003 the US Congress appropriated over $700 million to fund the National Nanotechnology Initiative. This is now a high profile body with links to many government agencies. It enables future research in the area and rolls out many educational programs. Nanotechnology is, not surprisingly, on the government's agenda. And this is not just an American phenomenon. In 2003, nine Korean government agencies, including the ministries of Information and Communication, Science and Technology and Commerce, Industry and Energy, have joined together to launch the $2 billion 2003 Nanotechnology Development Program.

Some civil rights groups have come out against nanotechnology stating that it will lead to technology that will invade the very core of what we are – think of the idea of the molecular level robot in your body unplugging your arteries – what would happen if some one, like a government agency, was to have veto over who received these robots and for how long. The power of technology is always a prerequisite of being in power. You might see Chapter 6 on wireless technology where we discussed some of the current day concerns of civil rights groups about RFID tags on products.

Whatever our concerns may be, researchers and scientists are going to do the work and try to come up with solutions to current nanotechnology issues. The possibility of using this technology far outweighs the downsides of not developing it in the first place. There is also the distinct possibility that if the US does not proceed with this research that someone else will. In the post 9-11 world this is not an option. Would it be better for a potential enemy of the US to have access to this potent technology? In fact, as with most nascent technologies, it is the arms and weapons manufacturers that are pushing nanotechnology research.

Another facet of nanotechnology that is intriguing is the area of replication. One of the early benefits identified for using nanotechnology was the idea that a very small scale, simple mechanical machine could be used to replicate itself. Molecular machines could be built that are minute versions of robots we find today in the automotive industry, for example. It is important to remember that we are considering molecular "machines" here. It is not intended to use nanotechnology to reproduce living organisms – that is a whole different ball game altogether.

Perhaps this idea of replication sounds daunting and perhaps even a little creepy. Our bodies are replicating – at a molecular level – every day so don't be too spooked. It is one of the goals of nanotechnology that replication will be able to offer the production of vast quantities of products that are useful to mankind, yet at a fraction of the cost. There will be very little pollution, wastage or expense. This is a desired goal of using this technology – the idea of self-assembly. This would, of course, be self-assembly of a pre-defined product using inputted materials, controlled programming and the relevant power supplied.

Let's take this one step further. Imagine we could program the replicator to make multiple copies of a molecular machine. Then, imagine we are able to program these machines to do other work – fastening small parts to larger parts and then have these larger parts perform other, slightly more complex work. Essentially we could build a whole product family, from the bottom up. This type of replication or assembly is known as convergent assembly.

A brief overview of some of the potential problems

Imagine we could build a robotic arm, shrunk down to fractions of a micron, which would be able to pick up and assemble other molecular parts – the subatomic version of working with nuts and bolts. These arms could be "programmed" to build a replica of themselves that would then be programmed to build another replica of itself.

In April 2000, Sun Microsystems founder and Chief Scientist, Bill Joy, hardly a technophobe, wrote a now famous article in Wired magazine about the negative side to this magical molecular mystery that may one day be ours. He argues that these nanotechnological auto-assemblers might get out of control and convert the planet and every living thing on it to a uniform but useless mass of bits and pieces: the **grey goo** (a term actually invented by Mr. Drexler). Mr. Joy goes as far as saying that there are some areas of research we ought not to pursue, because the consequences might be so dire.

Even the Unabomber, Ted Kaczynski, had some interesting ideas on the concept of nanomachines taking over:

"Eventually a stage may be reached at which the decisions necessary to keep the system running will be so complex that human beings will be incapable of making them intelligently. At that stage the machines will be in effective control. People won't be able to just turn the machines off, because they will be so dependent on them that turning them off would amount to suicide."

[Ted Kaczynski's Unabomber Manifesto, 1996]

Some more quotes on the subject:

The nightmare is that combined with genetic materials and thereby self-replicating, these nanobots would be able to multiply themselves into a "gray goo" that could outperform photosynthesis and usurp the entire biosphere, including all edible plants and animals.

[*American Spectator*, Feb. 2001]

Grey goo is a wonderful and totally imaginary feature of some dystopian sci-fi future in which nanotechnology runs riot, and microscopic earth-munching machines escape from a laboratory to eat the world out from under our feet.

[*Guardian*, July 2001]

Does this sound just a little bit scary to you – perhaps you are thinking that this sounds just like the way a virus – biological or software-based – functions. Nanotechnology is not designed to replicate these complex systems. The whole beauty of it is the way it could be made to replicate mechanical systems at a molecular level. Remember a mechanical system can only be as good as the inputs into it. So, for a molecular system to work we would have to feed it instructions, power and raw materials – without any of these it would not function. This is not true of biological systems.

One of the most advanced systems in nanotechnology being worked on is the example about the robotic arm mentioned above. It still requires controlling signals, power and raw materials. Some of the signals need to be delivered via a lithographic process. The control is still very much with the manufacturing system. There is also the Foresight Institute, based in Palo-Alto, which has written a set of guidelines for the principled development of molecular manufacturing systems. As mentioned above about organic replication, the Institute goes some way to encourage that this could never happen – replication should not take place in an uncontrolled environment.

What to Watch for in the Future:

This may be simplistic, but the thing to watch out for in the future is Nanotechnology itself. Nanotechnology, if the vision is realized, will effect everything. No one working in this area would deny that research is at an early stage and progress will be slow. There is some evidence that some of the issues we have discussed will come to fruition within the next decade, however.

The real breakthroughs in the use of nanotechnology will most likely occur in the medical and defense arenas. The idea of robots being programmed to routinely unclog your arteries is not that far-fetched either. Devices are currently available that can travel into an artery, take pictures and perform some unblocking functions as well. This obviously necessitates a visit to the hospital. Imagine you could have this device inserted in your vein, knowing that it will perform this artery cleansing work for ten years, based on its own pre-programmed workload and its own power supply. It would then simply dissolve harmlessly in the bloodstream and be replaced by a newer model.

In the areas of aerospace and weaponry there is a lot of activity, mainly in the materials area. We all know that diamond is a very strong material, due mainly to its molecular structure. It is feasible, using nanotechnology, to structure a stronger material out of covalent bonds by the more astute placement of atoms. This material could weigh less, be more malleable and 200 – 300 times stronger than diamond. Think of how useful this would be in aircraft design. The material could also be "intelligent" in that it can be controlled to bend or react instantly in certain circumstances – for example altering the structure of a wing during take off, landing and turbulence.

It is not an accident that NASA has people working very actively in the nanotechnology area. There is a huge potential benefit for space travel if new super-strong, lightweight materials could be found to build spacecraft. Not only could the craft be better designed but the spacesuits worn by the astronauts could be improved markedly.

Timeline:

The concept of Nanotechnology was first alluded to by the acclaimed physicist, Richard Feynman, in a seminal lecture given in 1959. As mentioned earlier, one of the next milestones in the development of the credibility of this area was in the groundbreaking work *Engines of Creation* by Eric Drexler, who invented the term nanotechnology and is the co-founder and Chairman of the Foresight Institute.

It is realistic to assume that nanotechnology will be part of our everyday lives in 15 years time. Nanotech devices will permeate every area of out lives – the clothes we wear, the batteries we use in our cell phones, the car we drive to work – and ultimately our longevity.

Researchers are coy about predicting when the first commercially available nano-machines will be produced but it will be some time yet. It is more probable that some developments will occur in the medical field that will herald the advance of nanotechnology into our daily lives.

Buzz Words and Definitions:

Bottom up- An approach to building things by combining smaller components, as opposed to carving them out of larger ones (top down), as is done in current photolithographic approaches to making silicon chips.

Convergent assembly- A process of fastening small parts to obtain larger parts, then fastening those to make still larger parts, and so on.

Diamondoid- Structures that resemble diamond in a broad sense, strong stiff structures containing dense, three dimensional networks of covalent bonds; diamondoid materials could be as much as 100 to 250 times as strong as titanium, and far lighter.

Fabricator- A small nano-robotic device that can use supplied chemicals to manufacture nanoscale products under external control.

MNT- An abbreviation for molecular nanotechnology that refers to the concept of building complicated machines out of precisely designed molecules.

Micron- One millionth of a meter, or about 1/25,000 of an inch.

Millimeter- One thousandth of a meter, or about 1/26 of an inch

Nanofactory- A self-contained macroscale manufacturing system, consisting of many systems feeding a convergent assembly system.

Nanometer- One billionth of a meter; approximately the width of a single strand of DNA.

Nanoscale- A scale significantly smaller than a micron.

Micro-Electro-Mechanical Systems (MEMS)- The integration of mechanical elements, sensors, actuators, and electronics on a common silicon substrate through micro-fabrication technology.

Nanotechnology

NNI- The US's National Nanotechnology Initiative.

Photolithography- The technique used to produce the silicon chips that make up modern-day computers.

Self-assembly- The process whereby components spontaneously organize into more complex objects.

The Bottom Line:

The bottom line about nanotechnology is that it is efficient, clean, inexpensive and will radically change the way we live. The fundamental difference between technology today and nanotechnology is purely a matter of size. Once we can get down to the level of building machines at the atomic and molecular range, the possibilities are endless.

Top Ten Cheat Sheet:

- Nanotechnology is about building machines at the nanometer level.
- A nanometer is a billionth of a meter, or about six atoms wide.
- Nanotechnology will be used in the aerospace industry to develop "intelligent" materials.
- Nanotechnology will be used in medicine to develop robots capable of doing tasks like cleaning out your arteries, on a routine basis.
- The US leads the way in this area; accounting for up to 30% of the total global research spent on nanotechnology.
- Nanotechnology has the possibility to revolutionize manufacturing processes – offering clean, efficient and cost effective systems.
- Nanotechnology will effectively replace current lithographic technologies, using an integrated circuit fabrication.

- The number of transistors on a chip will be massively increased from current day levels.

- Nanotechnology will not be a biological process but will focus on producing machines that can replicate in a controlled environment.

- Self-assembly systems can be built that will replicate, at the very small scale level, complex mechanical processes.

3

Security

"If I were two-faced, would I be wearing this one?"

- Abraham Lincoln (1809-1865)

Throughout this book we have been discussing new technologies that will change the way we do business and the way we live our lives. Incumbent with all technology is the ability to successfully achieve a certain task. This success can often be compromised by system failures or system attacks. This chapter deals with how systems are attacked, or compromised deliberately, and how the technology fights back. We have all been exposed to security issues in our technological lives – whether it is accessing a network in the office, running a local password protect on a Word document or getting viruses and worms delivered to our inboxes.

Let's look at the one that is of immediate concern to all of us - Spam - and the possibility of having our computer corrupted by viruses contained within these unsolicited messages. We all know what to do when we receive these annoying messages – into the trash or junk mail folder. It is estimated that we spend over have of the email time we spend each day dealing with spam. On its own spam is merely an annoyance and prudent use of Outlook tools can preclude most of it getting near you. The real issue with spam or unsolicited email is that it is the most frequent channel for spreading viruses and worms to your machine.

Put simply, a computer virus is a self-replicating program that explicitly copies itself and that can infect other programs by modifying them. In the early days of virus propagation this involved having to run an executable (.exe file) attached to an email. Then viruses become cleverer and you only had to open the email to become infected. Then along came worms. A worm is a self-contained program that breaks into a system via remotely exploitable security flaws and self-replicates without requiring an external action, like the user running a program or re-booting their system. One such potent example in the Nimda worm. Nimda spreads automatically via shared files, web pages, email and other routes. Infected computers can be cleaned, but this worm spreads with such speed and in such volume that networks can grind to a halt. We will examine the Nimda worm in the Detailed Summary of this chapter.

Security experts admit such attacks cannot be prevented entirely, the key being to ensure that your network is protected by an up to date security and firewall package. Vendors of security protection software say simulation technologies now in development might at least help network operators predict how their systems will respond to invaders, so they can prepare better defenses and contain the damage.

To show that the Government is taking the issue seriously, the Bush administration has passed a Federal law to combat the spammers. This is known as CANSpam and involves penalizing spammers to the tune of $250 per email address used. While this might sound like an excellent measure to combat spam it is worth noting that it is US Federal Law and so has no jurisdiction outside the US – and many of the spammers can locate their servers offshore and be immune from such penalties. However, it does send out the right signals and if Europe and Asia develop similar laws it will become more difficult for the spammers to ply their trade.

There is also the government funded CERT Coordination Centre, a research and development center for internet security, based at Carnegie Mellon University. Security experts there are working to remedy individual vulnerabilities, but they agree the virus makers will always be able to find new ways to intrude. They have developed Easel, a software simulation tool that runs potential nightmare scenarios involving the likes of Code Red and Nimda. Using the collected data from previous attacks – how many servers were affected in what span of time, for instance – it creates reference models that computer security specialists can use to minimize damage in future attacks. They might, for example, configure a network to recognize a nascent infection and shut down affected servers before the virus can spread further.

Similar software is under development at security software giants such as McAfee and Symantec. The emphasis seems to be shifting to remedy rather

Security

than prevention – there will always be spammers and virus spreaders – the key is developing ways to deal with these occurrences.

Detailed Summary:

Some of the main email providers are actually trying to nip spam in the bud to stop it at its source, so to speak. Two popular web-based email services - Yahoo! and Microsoft's Hotmail - employ "tricks" to prevent spammers from automatically signing up for hundreds of mail accounts that can then be used as spam launch pads. These "tricks" are also known picturesquely as "captchas" which actually stands for "completely automated public Turing test to tell computers and humans apart." Phew! That's a mouthful – the Turing part refers to computer genius Alan Turing who once said that a computer could only be truly intelligent if it could fully emulate a human being. These captchas are of great assistance because most spam is being generated by robots or automated software senders.

You may have seen these captchas when you have signed up for an email service. It asks you to enter in a word from a distorted picture of the word or sequence of numbers. Their automated software cannot recognize the code in the same way that a human can and so it can block these auto sign-ups. Search engines are also using this technique to stop multiple URL submits from auto-submission software – which can clog up their servers and effectively slow the system down.

Spam Arrest, a junk email blocking service, uses the captcha technique to filter out machine-generated email. The service is based on the ability to visually recognize words. Try sending an email to a Spam Arrest user, and you might be presented with a fuzzy word against a complex and distracting background. Your brain will be able to decipher the word, no problem, but computer recognition programs would have difficulty in completing this simple task.

The Palo Alto Research Center have taken this approach one step further. They devised a system that generates words that may sound English but are misspelled and have chunks taken out of the letter, set against a fuzzy visual background. This obviates the possibility of dictionary generating programs whizzing through all possible word combinations. This approach could stop spammers from creating accounts that send out the junk email.

Another invidious technique used by computer saboteurs is to swamp a company's web server with bogus internet traffic. Some of the world's biggest websites, from a traffic point of view, have been attacked in this manner, including eBay, Amazon.com, and Microsoft's Hotmail and Expedia. The saboteurs do not need to breach the security of the target network. The trouble with this type of attack is manifold; there is no sophisticated software needed, the traffic looks like normal internet traffic (to a point) and the sheer volume of the attack may produce what is known as a "denial of service" scenario – basically the server is swamped and not available to you and me to access as normal.

So how does a company go about identifying this "bad traffic"? One company offering a solution to this problem is Arbor Networks, based in MA. Their software helps to identify what the typical traffic profile is and to flag any deviation from this. Thus, any suspicious fluctuations to normal traffic can be flagged. The software can tell the difference between "normal" surge traffic generated from an intensive advertising campaign because of the nature of how this saboteur traffic originates, tending to come from a small number of machines making thousands of data requests. The bad traffic can then be relatively easily filtered out from the system.

The Network Problem

You may access the internet and your email from the office server. You can guarantee that you are doing this from behind a firewall. Basically, a firewall filters all network traffic to determine whether to forward the traffic towards its destination. A firewall is effectively a buffer between the "outside" and your network, programmed with rules that determine what traffic is allowed to pass and what is to be blocked. For example, a firewall might be set up to allow managers in technical resources to browse the internet, or to remotely access their PCs from home, while only permitting people in marketing to access their email. Most firewalls, no matter what their level of access, record and log everything that traverses the server boundary and can give a quick snapshot of what the traffic profile is for the company.

It is possible to get both hardware and software firewalls. There are a number of firewall screening methods. At their simplest level these devices screen data requests to make sure they come from acceptable domain names and IP addresses. If it is necessary for users to access the system remotely, for an on the road sales rep, for example, then a secure and authenticated logon can be provided. One problem with using a firewall is that they can lure a company or organization into a false sense of security. Firewalls do provide a degree of security to an organization but it is really just perimeter security.

Security

There is still the threat of a security attack from within the perimeter or the perimeter being breached. They also have the downside of making internet access slower.

The reason organizations still opt for firewalls is that they make life a lot easier for the IT department. Imagine that a new fix is needed for a particular virus or bug. The IT department might have to update thousands of PC registers instead of simply bolstering the firewall. Unfortunately, a lot of cases of security breach come from within – employees using laptops outside the network and then reconnecting without adhering to security processes or disgruntled employees made redundant or with a score to settle with an old boss! Many companies who succumbed to worms like Code Red and Nimda believed they were fortified securely behind their firewall. Many an IT manager might have faced stiff questioning over these issues. Let us take a look at one infamous worm – Nimda – in a little more detail.

The Nimda mass-mailing worm first manifested itself in 2001, thanks to a backdoor left by the Code Red II worm. The somewhat exotic name of the virus came from the reversed spelling of the very mundane word "admin." What was important (and potent) about this worm is that it was the first worm to propagate itself by a number of methods, including email, network shares and an infected website. At the time it was unleashed Computer Associates rated it at number eight in its Top Ten Virus threats List.

This worm sends itself out by email, searches for open network shares, attempts to copy itself to unpatched or vulnerable Microsoft IIS web servers, and is a virus infecting both local files and files on remote network shares. When Nimda was first unleashed the traffic it generated caused many "denial of service" occurrences.

The real invidious part of Nimda is that it compromises the security of the host it infects and lets a remote user have full administrator authority and access to all the system files. As if this is not enough to contend with, Nimda makes numerous changes to system files, which make it notoriously difficult to clean. To detect this worm was serious enough and required a lot of IT time. The task of cleaning it required some patches, software updates and potentially some server downtime. Even with the cleaning operation recommended by most software vendors, there was always the possibility that more backdoors, left by Code Red or another worm, would be lurking on the network waiting for the next strike.

The use of firewalls in the Nimda scenario was clearly inadequate. However, that is not to say that firewalls are useless. They are extremely important and are a definite necessity. The problem is that IT departments think that they are completely protected once the magical firewall is installed. The firewall is a perimeter defense and we have mentioned that it is only as good as the other defenses: regular system backups, suitable level encryption and even checks on what employees are doing with their surfing time. Remember the disgruntled employee we mentioned earlier – well try to make sure that this person has limited access to sensitive IT resources. It may sound draconian but it is necessary to ensure the integrity of your systems. Take it that all of your systems are already insecure, from the moment they are activated, so installing firewalls, performing regular backups and keeping one eye on employee activity is a prudent and strategic way to keep your network secure.

Wireless Security

In Chapter 6 we will discuss various wireless technologies – RF, Bluetooth, Wi-Fi and GSM - that you'll need to concern yourself with. It is interesting in this chapter to look at some of the security issues relating to these technologies.

It is reasonable to say that we should not concern ourselves too much with security on our cell phones. If you use an analogue cell phone anyone with a decent scanner can tune in to what you are saying – no way around that. But GSM phones use encrypted technology that enables secure voice and date transfer during calls. GSM technology uses an algorithm to ensure the authenticity of the caller and the integrity of the channel, even when you are roaming in a foreign country.

The most interesting, and potentially contentious area of wireless security is that concerning wireless LANs or Wi-Fi networks. As will be alluded to in Chapter 6 these are fast becoming the connection method of choice, whether in the humble home office environment, over a coffee in your local Starbucks or in a corporate office. It is a regular occurrence that someone is sitting in an office or a hotel in a metropolitan area and picks up a network signal of a local wireless network. Wireless signals do not recognize corporate or geographical boundaries and are only limited by the propagation configuration of the network. Even in an office environment you will find small areas or "blind spots" where the coverage is very weak or non-existent. So, it is possible for the random surfer to "happen upon" or

Security

"jockey-back" on someone else's network. So how do you protect against this happening?

As you will see in Chapter 6 wireless local area networks use spread-spectrum technology - a technique that makes the radio signals difficult to intercept. Most Wi-Fi systems also include a form of user logon and password protection. Of course, the spread spectrum signals can be intercepted with a relatively simple wireless card and many networks do not properly set up the password feature and will allow ready access to someone typing the word "any" as a password. The fact that "employees" have to go through some form of physical security before they can access the network only adds to the notion that wireless networks may not be as secure as equipment manufacturers would have us believe.

The problem with wireless security is essentially a technical issue with the way the signals are encrypted. The original wireless LANs (WLANs) used the Wireless Encryption Protocol (WEP). This was then replaced in late 2002 with the Wi-Fi Protected Access (WPA). Essentially, WPA offered improved data encryption through the use of temporal key integrity protocol (TKIP). The TKIP feature scrambles the keys using a hashing algorithm and ensures that the keys have not been tampered with. WEP only uses a static key that is seldom changed by users. This cryptographic weakness caused many of the security breaches in WLANs because intruders could, with relative ease, generate an encryption key and access a wireless network.

While WPA offers enhanced security features over WEP, not all industry observers are completely satisfied. A recent problem was highlighted with WPA concerning the use of poorly chosen passwords for a network. Criminals intent on compromising a WLAN can use simple dictionary software to overcome the system password. In fairness, this weakness only manifests itself when short, text-based keys are used and does not signify a fault in the WPA protocol. WLAN manufacturers can circumvent this problem by incorporating the ability to generate random keys across the network and putting in place user requirements concerning the length and style of passwords.

Microsoft responded to this potential threat by providing a Windows XP download that alters the way the operating systems communicates with the Wi-Fi network – using separately generated keys for each system user rather than one, albeit encrypted, key for the network connection.

A Big Brother State

Some of the encryption, security and signal encoding issues that we might blame the manufacturers on may in fact have another source of culpability. During the 1990s the US Government was urging industry to pursue a policy of "weak encryption." The government's argument was based on the belief that building highly encrypted, absolutely secure products would be an impediment to their intelligence gathering operations! Cryptographers argue that they can now make encryption systems that are virtually unbreakable. Here is an excellent illustration of how difficult it can become to crack a moderately encrypted signal. A small bank of Pentium enabled computers can search a billion keys a second and can crack a 40-bit encryption key in 18 minutes. Double the key length to 80 bits and those computers would have to grind away at the math for a little while longer - about 38 million years to be precise. With commonplace and easy to generate 128-bit encryption it would take billions of these computers trillions of years to crack! No one is going to wait around that long to see it happen – the Sun will have died and our little planet with it long before.

It is unclear who is winning the argument for tougher security. In the post 9/11 world the government is in a strong position to browbeat security wireless system manufacturers to abide by their "guidance." It is interesting to note that Bluetooth devices (see Chapter 6) base their PIN security on a 14-bit encryption system – note the mind boggling numbers to crunch here. While there is authentication and random number generation built into the standard, the system is only as secure as its weakest link. Bluetooth is excellent for connecting wireless device together in a short-range environment – wireless mouse and wireless keyboard, cell phone to PDA – but it may not be secure enough to transfer money or sensitive information across.

Security

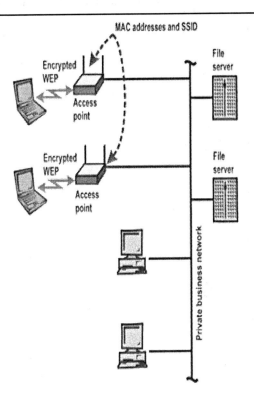

Figure 3.1: Wi-Fi Network Security

What to Watch for in the Future:

You should expect that security aspects of all areas of technology will be radically improved upon. This is a direct response to the perceived threat of cyber crime and that an equivalent 9/11 could be unleashed on the Western World in the form of a virulent virus, worm or internet attack. Systems and networks will have to become much more secure. Microsoft has taken a tentative step and has openly declared war on spammers. The US and European administrations are working aggressively to combat spam and its associated viruses and worms. It is possible that your email will become a slightly more time consuming and cumbersome method of communicating as many levels of checks and cross checks are part of the send and receive process. This will be a small price to pay if the integrity and usability of your email can be improved.

Another interesting area to watch out for is called Quantum Cryptography. With conventional cryptography the idea is to build the strongest locks and bolts to safeguard information. However, this is only secure so long as no one steals the key! Quantum cryptography ensures that the key is secure as well. It uses individual photons (packets of light energy – remember Physics 101) for key transmission and polarizes their direction randomly. The beauty is that any attempt to tap into this signal alters the polarization in a way that both the sender and receiver can measure – so they know the system has been compromised (or an attempt has been made to compromise the system). They will then continue to transmit, using different keys until a non-disturbed key gets through. This is an exciting area with many of the big guns, including IBM, giving it serious attention.

Buzz Words and Definitions:

Backdoor- A piece of software that allows for unauthorized access to a system.

Firewall- A firewall is a set of related programs, located on a network gateway server that protects the resources of a private network from users from other networks. Basically, a firewall, working closely with a router program, filters all network packets to determine whether to forward them toward their destination.

Malware - Malware stands for Malicious software. A catch-all term for 'programs that do bad or unwanted things.' Generally, viruses, worms and Trojans will all be classed as malware, but several other types of programs may also be included under the term.

Mass Mailer - A virus that distributes itself via email to multiple addressees at once is known as a mass mailer. Probably the first mass mailer was the CHRISTMA EXEC worm of December 1987.

Virus - A computer virus is a self-replicating program that explicitly copies itself and that can infect other programs by modifying them. It can also modify their environment so that a call to an infected program implies a call to a possibly evolved copy of the virus.

Worm - Self-contained programs that break into a system via remotely exploitable security flaws. They then self-instantiate - their replication

Security

mechanism is directly responsible for their code running on new target host systems, rather than requiring some external action such as a user running a program or restarting the system as with viruses.

The Bottom Line:

If your organization does not take all forms of security seriously – email, internet, network, wireless LAN – it is not a matter of if you will be seriously compromised – it is only a matter of when.

Top Ten Cheat Sheet: Security

- Email now comprises over 50% spam.

- Spam is a massive network overhead and serious efforts are being made to route it out at its source.

- Email security can be as simple as using Outlook to examine suspect incoming email and moving it to a Junk folder to having a corporate wide firewall in place.

- Firewalls are not a security panacea.

- Firewalls are only as good as the other security process – regular backups, employee vetting and regular network security upgrades.

- Wireless security poses different issues because of open nature of access to it.

- Wireless LANs must be made as secure as possible using all techniques applicable to cabled networks – and then some.

- Security will become a more important technology for your organization over the next decade.

- Internet security might involve slowing access to your staff but will be outweighed by reduction in system downtime and improved immunity to viruses and worms.

- Advances in cryptography, including Quantum Cryptography, will enable completely secure key transmission to take place.

4

Grid Computing

"I can write better than anybody who can write faster, and I can write faster than anybody who can write better."

- A. J. Liebling (1904-1963)

Grid computing is all about using the untapped resources of computers connected to a network. IBM perhaps defines it most succinctly: Grid computing is applying resources from many computers in a network-at the same time-to a single problem.

Remember the idea of connecting lots of computers together is nothing new – just think what the Internet does for document sharing. But grid computing holds out the holy grail of using the actual unused computing power of all the machines connected to the grid to perform some high-end computations. The technique of using distributed processing power, and distributed storage and access, became popular through schemes to use the unemployed processing power of connected machines around the world.

An example is the SETI@Home project (www.seti.org), which uses the processing power of thousands of connected home computers to search for extraterrestrial life in the stars. SETI stands for "Search for Extraterrestrial Intelligence" and is a private, nonprofit organization dedicated to scientific research, education and public outreach. Volunteers can participate by downloading a free screen saver program that analyses chunks of radio telescope data on their PC when it is not in use and communicates this data back to a central server via the Internet. This project could never have been achieved with the humble processing power of even the most advanced supercomputer. Another example is the unraveling of the SARS virus – this was achieved using the power of grid computing.

In a business environment, partly for security reasons, grid computing is mainly represented by grid systems within enterprises, rather than from unknown sources, to better utilize available processing power. The real benefit of a company using a grid environment is that computing power can be purchased in an on demand manner, much the same way as other utilities like electricity or water. Thus, a company can happily use its own computer infrastructure for 11 months of the year and then buy-in extra computing power for one month for a particularly heavy sales period, for example. A company like Amazon.com relies heavily on its computer infrastructure. During the year there are peaks on its requirements – extra searches and sales during holiday periods – and it will be able to purchase extra resources on an "as needs" basis. This allows a company to invest more accurately and efficiently in technology.

Detailed Summary:

The development of the World Wide Web revolutionized the way we think about and access information. We really don't think twice anymore about logging on to the web and pulling up information on almost any topic imaginable. What the Web did for information, Grid computing aims to do for computation. Grid computing is really the next logical evolution of the Internet.

The Internet began with TCP/IP and networking; then came communication with e-mail, followed by information sharing with the World Wide Web. Next will be the advent of grid computing, the sharing of actual computer resources, such as memory, storage, and processing power.

It is almost mind boggling to imagine the types of applications that could be developed if access to distributed supercomputers, mass storage and vast memory were as straightforward as access to the web. So there are several ways of looking at Grid Computing: as a way to connect the computational power of all the big computers together and give access to companies and academia alike; as a way to connect ALL the computers both big and small and derive computational efficiencies (think peer-to-peer networks); as the next logical step in providing a computational platform for Web Services (see Chapter 1); as a business on demand or computing as a utility model which IBM and others are touting.

All of the above are in effect true. Through a variety of different means and technologies, computers will learn to share each other's processors, storage and memory, much as they share communications and information today, and applications will take advantage of these resources. While we are far from realizing the full effects that Grid computing will bring, it IS upon us and there are both short-term and long-term ramifications for the enterprise.

The actual brains of the computers will be connected, not just the arteries. This means that users will begin to experience the Internet as a seamless computational universe. Software applications, database sessions, and video and audio streams will be reborn as services that live in cyberspace.

Once plugged into the grid, a desktop machine will draw computational power from all the other machines in the grid. The Internet itself will become a computing platform. Grid computing is the next logical step for the Internet to take.

Imagine the Web as a high-speed I/O (input/output) bus for large-scale distributed computers. The processing power could be in one facility, the memory in another and the storage in yet another all connected by gigabit pipes.

In August 2001, the National Science Foundation (the same organization essentially responsible for the Internet) began the first steps in launching the TeraGrid, a transcontinental supercomputer that should do for computing power what the Internet did for documents. TeraGrid is a multi-year effort to build and deploy the world's largest, fastest, distributed infrastructure for open scientific research. The TeraGrid project was launched with four sites: the National Center for Supercomputing Applications (NCSA) at the University of Illinois; Urbana-Champaign, the San Diego Supercomputer Center (SDSC) at the University of California, San Diego; Argonne National Laboratory in Argonne, IL; and Center for Advanced Computing Research (CACR) at the California Institute of Technology in Pasadena.

In October 2002, the Pittsburgh Supercomputing Center (PSC) at Carnegie Mellon University and the University of Pittsburgh joined the TeraGrid as a major new partner when NSF announced $35 million in supplementary funding.

Another $10 million in NSF awards in 2003 will add four additional sites to the partnership: Oak Ridge National Laboratory (ORNL), Oak Ridge, TN; Purdue University, West Lafayette, IN; Indiana University, Bloomington; and the Texas Advanced Computing Center (TACC) at The University of Texas at Austin. Primary corporate partners are IBM, Intel Corporation, and Qwest Communications. Other partners are Myricom, Sun Microsystems, Hewlett-Packard Company, and Oracle Corporation.

When completed, the TeraGrid will include 20 teraflops (that's 10 to the power of 12, a trillion, floating point operations!) of computing power distributed at five sites, facilities capable of managing and storing nearly 1 petabyte of data, high-resolution visualization environments, and toolkits for grid computing. These components will be tightly integrated and connected through a network that will operate at 40 gigabits per second, effectively forming the fastest research network on the planet.

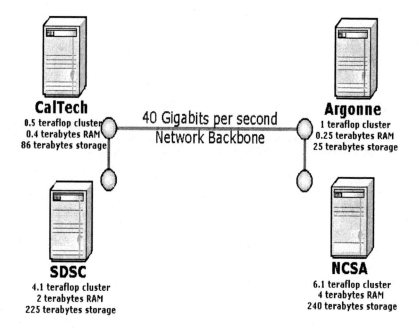

Figure 4.1: Initial Four TeraGrid sites

This virtual computer will rip through problems at up to 20 trillion floating-point operations per second, or teraflops – eight times faster than the most powerful academic supercomputer available today. Such speed will enable scientists to tackle some of the most computationally intensive tasks on the research docket – from problems in protein folding that will form the basis

Grid Computing

for new drug designs to climate modeling to deducing the content and behavior of the cosmos from astronomical data.

Computing grid power will be like the electrical grid: Just as you plug in a hairdryer, you will plug in your computing device and pay for the megahertz and megabits you use. This gives rise to the concept of computing as a utility for e-business applications.

Irving Wladawsky-Berger, vice president, Technology and Strategy, IBM Server Group, calls grid computing the "key to advancing e-business into the future and the next step in the evolution of the Internet toward a true computing platform." He predicts that grid computing, like supercomputing before it, will find its way into the commercial world and provide a vast infrastructure for e-business.

A recent advertisement from IBM illustrates how some of the infrastructure players see grid computing developing:

"E-business on demand™ is The Next Utility™."

Thirsty? Turn on the water tap.

In the dark? Flick on the light switch.

Looking for e-business solutions? E-business on demand makes them just about as easy as working with any other utility.

Grid computing is data storage, Web hosting, e-procurement, and more – a wide range of e-business services managed, hosted, serviced, upgraded, and delivered to your door – and you don't have to worry over the details. With e-business on demand, end-to-end e-business can be as accessible, affordable, and easy to use as water, gas, telephone, or electricity.

By giving up the burden of IT ownership, your company can vastly increase its access to computing power, expertise, and innovation. Subscribe to what you need today. Turn it up when you need more.

For large companies, grid computing will save millions in up-front investments in hardware, software, integration, and hiring. For growing companies, grid computing gives them the sophisticated network of a larger competitor at a fraction of the cost.

With e-business on demand, we focus on your technology; you focus on your business. Together, we focus on your success."

It seems to be working. American Express has recently signed a seven-year contract with IBM Global Services to outsource its entire IT infrastructure, buying IT capacity and capabilities from IBM on a utility basis. The $4 billion deal is the largest ever for IBM's utility computing concept, which charges companies according to how much computing power or other IT they use, according to IBM Global Services financial services general manager Paul Sweeny. He says American Express foresees "hundreds of millions of dollars" in savings through the contract, and that the utility-based model was even more attractive, given the current recession.

Grid computing has been primarily a tool for large computationally intensive R&D and scientific applications with current deployments falling mainly in the areas such as the life sciences, financial services and scientific research. However, as with examples like American Express above, Grid is entering the IT mainstream.

One of the keys for moving grid computing forward will be the creation and acceptance of a set of standards and protocols. The Internet and Web would never have gotten off the ground if we hadn't agreed to and accepted proctors such as TCP/IP, HTTP, and HTML. Many of the Internet protocols we use today came from the research community – the same people who are pushing for grid computing standards.

IBM appears committed to promoting open source protocols for grid computing. For the short-term, their strategy for grid computing is to make all servers grid-enabled, that is, implementing a common protocol. As such, IBM is working with outside groups, such as the Global Grid Forum and Globus. Jointly, IBM and Globus have submitted a specification called the Open Grid Services Architecture (OGSA) to merge grid computing protocols with Web service protocols and standards. Additional key considerations about grid computing technology will include security and resource management.

Their grid computing strategy aims to develop protocols to link servers and storage systems to combine their processing power. Interestingly, IBM has developed these protocols in conjunction with Ian Foster and Carl Kesselman, the scientists accredited with developing the concept of grid computing.

The really important thing about this announcement is that the protocols will be open and not IBM's. To reinforce this idea Linux will be at the heart

Grid Computing

of this work, ensuring that on the computing grid the right power is allocated to the right computer users wherever they may be on the grid.

IBM is also positioning itself to take the power of grid computing into the enterprise. In line with security issues it is envisaged that enterprises will set up what are known as 'intragrids' – effectively keeping all the computing and information gathering within a single enterprise. This is still grid computing – with all its attendant cost efficiencies – but it is not opening up a model for ALL computers to connect to a grid (like the SETI@Home project alluded to earlier).

Further down the line IBM will develop grids designed to suit specific industries, including petrochemical, financial services, media, telecoms and others. These will include availability grids from service providers and utility grids offering storage by IT services companies. To ensure this gets buy-in, the open source route is the only way to proceed for IBM.

While the full ramifications of grid computing, or utility computing, may not be realized for a long time yet, when giants like IBM start talking up the game we know it will not be too long before we can reap the benefits.

One of the bodies to the forefront of Grid standards development is the Globus Alliance. The Globus Alliance (http://www.globus.org) is a research and development project focused on enabling the application of Grid concepts to scientific and engineering computing. In case you think this is purely a US organization then think again - the Alliance is a partnership of Argonne National Laboratory's Mathematics and Computer Science Division, the University of Southern California's Information Sciences Institute, the University of Chicago's Distributed Systems Laboratory, the University of Edinburgh in Scotland, and the Swedish Center for Parallel Computers. Grid computing is global.

The Globus Alliance have developed the Globus Toolkit - effectively a set of useful components that can be used either independently or together to develop useful grid applications and programming tools.

Applications that can benefit from Globus technologies can be loosely classified as one of the following types:

- **Distributed Supercomputing:** These applications have large computational requirements that can be met only by simultaneous execution across multiple supercomputers. An example would be NASA's OVERFLOW-D2 application, which simulates airflow around an airborne vehicle.

- **Smart Instruments:** Computational grids can enhance the power of scientific instruments by providing access to data archives and on-line processing capabilities. The X-Ray CMT application uses Globus to couple Argonne's Advanced Photon Source to a supercomputer in order to perform sophisticated on-line processing and interactive visualization.

- **Medical Research:** The Smallpox Research Grid Project, which combines the efforts of IBM, United Devices and Accelrys, is helping to feed leading screening researchers at Oxford and Essex Universities in the U.K. as well as smallpox experts at the Robarts Research Institute and Sloan Kettering Cancer Center with the computing resources they need to help identify new anti-viral drugs. When the 2 million PCs are linked, they essentially serve as a virtual supercomputer capable of delivering 1,100 teraflops at peak performance, or about 30 times the power of the fastest supercomputer currently available.

How will Grid Computing Affect Businesses?

Grid computing will open up a whole new horizon of business opportunities. The sheer power available and the OnDemand delivery approach will radically reshape the way an enterprise configures its IT infrastructure. Below are some examples of how this technology will affect your business:

- Teraflop desktops: Chemical modeling, symbolic algebra, and other packages that transfer computationally intensive operations to more capable remote resources.

- Collaborative engineering (aka teleimmersion): High-bandwidth access to shared virtual spaces that support interactive manipulation of shared datasets and steering of sophisticated simulations, for collaborative design of complex systems.

Grid Computing 45

- Distributed supercomputing: Ultra-large virtual supercomputers constructed to solve problems too large to fit on any single computer.

- Parameter studies: Rapid, large-scale parametric studies, in which a single program is run many times in order to explore a multidimensional parameter space.

How will Grid Computing Affect Consumers?

The most tangible effects of grid computing will be in realizable benefits enjoyed by the consumer. The availability of cheaper drugs and more enhanced computing power at the local desktop is a real goal of this technology. Consumers and the public can also participate in global projects like SETI@Home and The Smallpox Research Grid Project by donating their unused processor time.

What to Watch for/Future:

As with several of the newer technological movements we have discussed, we invariably come back to the question of standards and Grid Computing is no different. While grid computing standards and technologies are maturing there are still more than 20 working groups in place, for example, to define things such as the OGSA (Open Grid Services Architecture) platform and security for grids.

Grid technologies are also gaining the backing of major vendors, including IBM, Hewlett-Packard, Oracle, NEC, Fujitsu, Avaki and Platform Computing. All of these vendors are rolling out support for the Globus Toolkit, which will encourage take-up of grid computing technologies. Grids are going to be a key enabler of new technologies and it is imperative that the major box manufacturers and systems integrators work towards the common standard approach. If this happens, then we will all see grid computing affecting every area of our lives.

Buzz Words and Definitions:

Floating Point Operations- This is a method of encoding real numbers within the limits of the finite precision available on computers. Using floating-point encoding, extremely long numbers can be handled relatively easily.

Globus Alliance- A collaborative academic project centered at Argonne National Laboratory focused on enabling the application of grid concepts to computing.

Globus Toolkit- The Globus Toolkit is an open source software toolkit used for building grids. The Globus Alliance and many others all over the world are developing it. A growing number of projects and companies are using the Globus Toolkit to unlock the potential of grids for their cause

Grand Challenge- A problem that by virtue of its degree of difficulty and the importance of its solution, both from a technical and societal point of view, becomes a focus of interest to a specific scientific community.

Grid computing- A type of distributed computing in which a wide-ranging network connects multiple computers whose resources can then be shared by all end-users; includes what is often called "peer-to-peer" computing.

Teraflop- A teraflop is a measure of computer processing power and is a trillion (10 to the power of 12) floating-point operations per second.

Terabyte- A terabyte is a measure of computer storage capacity and is 2 to the 40th power or approximately a thousand billion bytes (that is, a thousand gigabytes).

Petabyte- This is a massive memory storage number equaling 2 to the power of 50 bytes – or 1,024 terabytes – or about 20 million four-drawer filing cabinets full of text documents!

The Bottom Line:

Grid computing will take many forms and be called many things from peer-to-peer networks to distributed computing but the one thing we do know is that it is here to stay. It will impact your work and business life. Grid computing will unleash massive computational power that will be used in medical, financial, research and business applications. The power of grid computing is many orders of magnitude above what the "humble" supercomputer can offer today and because the component of this new supercomputing power doesn't have to be in the same geographical space the possibilities will be endless.

Grid Computing

 Top Ten Cheat Sheet: Grid Computing

- Grid computing combines the computational power of many computers connected over a network
- Grid computing enables processing rates in the order of teraflops
- A teraflop is 10 to the power of 12 floating point operations per second
- Grid computing will be available to an enterprise in an OnDemand format
- Grid computing will enable the home user to participate in global, social projects aimed at solving scientific or medical problems
- Grid computing will enable your organization to make cost savings on IT infrastructure
- Grid computing will enable your organization to tap into massive computing power, if and when it is required
- Grid computing will radically alter the way computational intensive processes are enacted
- Grid computing is supported by the major global payers like IBM and Sun
- Grid computing standards are Open and under the aegis of the Globus Alliance

5

Linux

"Imagination is more important than knowledge"

- Albert Einstein

Linux is a UNIX like operating system that was first developed in 1991 by a 21-year-old Computer Science student at the University of Helsinki named Linus Torvalds. Linus made his creation and code available on the Internet so hundreds and then thousands of avid programmers could tinker with and improve his creation. At that time the main operating systems in place were Microsoft DOS for PC users, UNIX for high-end business users and the Apple Mac OS. The one thing these Operating systems had in common was that they were proprietary, meaning regular users couldn't gain access to the underlying source code. (While UNIX started out with an open code base distributed by Bell Labs, it has since been commandeered and privatized by the big technology companies.)

In an attempt to create an operating system that could help students understand the inner workings of a computer, a Dutch professor named Andrew Tannebaum created MINIX, a 12,000 line operating system written in C programming language and assembly language. MINIX became very popular amongst computer science majors around the world, eager to get a glimpse at how an operating system works. One of those students was Linus Torvalds. Linus thought MINIX was good but he thought he could do better.

Inspired by the GNU PROJECT 1984 by Richard Stallman, a movement to provide free and quality software (the beginnings of what we know today as the Open Source movement), Linus determined that he would make his code available via Internet Newsgroups (this was before the World Wide Web).

So he posted to a MINIX newsgroup and announced his plans, looking for volunteers in need of a meaty project, to help him tinker with and improve his new Operating system. They responded, in the tens, then hundreds, growing eventually to hundreds of thousands of programmers worldwide all working on different parts of Linux and writing applications. This was, and still is, a modern marvel. A young man created a software program and leveraged the power of the Internet and the open source software movement to create something that has changed and continues to change the very face of computing, as we know it.

Detailed Summary:

One of the biggest worries for the Linux community is "Will Linux become UNIXed?"

What I mean is this: Unix began life as an Open Source operating system freely available for everybody to view its code and do what they liked with it. Over time, though, UNIX was bent into over 31 different proprietary versions by Companies like IBM, Sun and Hewlett Packard to name but a few. Software applications would run on one version but not on another. This is one of the main reasons there was a big opening for an Operating system like Microsoft Windows NT (the enterprise version of Windows) – the same programs ran on all installations of NT, there is a standard platform. Ironically that is one of the same reasons Linux is getting popular today; it has so far remained standardized under the auspices of the GNU license and oversight. Thus, the fear is that as Linux becomes more and more popular some of the larger software and hardware manufacturers will start creating their own proprietary versions of Linux for their own special interests. We might even see a Microsoft Linux one day!!

But I believe for that very reason companies like IBM, DELL, Oracle, HP and SUN will not release their own versions of Linux, in the foreseeable future at least. It is in all their best interests to create a strong alternative to Microsoft Windows and the best way to do that for the moment is to create a strong and standardized Linux Platform and continue to push the tremendous growth Linux has seen.

Linux is gaining momentum in nearly all corners of computing from clustered supercomputers to PDAs to desktops. In fact, large portions of this book were written on a Linux powered device.

Linux

Wal-Mart is currently selling low-end PCs running a Linux OSes called Lycoris and Lindows.

Sharp has now released its third version of its Zaurus PDA running Linux.

IBM is testing a Linux based watch. Furthering the reach and application of Linux are projects like the TeraGrid. Launched by the National Science Foundation (NSF) in August 2001, the TeraGrid project is a multi-year effort to build and deploy the world's largest, most comprehensive, distributed infrastructure for open scientific research, capable of performing over 20 trillion calculations per second (20 Teraflops) by 2004 (See Chapter 4 on Grid Computing). At each of these centers, there is a supercomputer. In total, there will be more than 3000 processors running in parallel to create the TeraGrid. The main technology behind this massive computer is clustering: the technology of binding together many low performance/cost processors to create a single computing environment.

Dubbed as the Distributed Terascale Facility, the TeraGrid will combine enough computing power to facilitate the solution of complex mathematical and simulation problems, ranging from Astronomy and Cancer Research to Weather Forecasting. Equipped with a 600 Terabyte storage space, the TeraGrid will be so powerful that it would take a human working on a calculator 10 million years to do what the TeraGrid can do in only 1 second. The amazing thing is that Linux will power large parts of this massively parallel Mega Computer. Each of the 4 sites mentioned above will operate a Linux cluster, and connect by means of a 40 Gigabit/sec dedicated optical network.

In May 2002, paint retailer Sherwin Williams rolled our Linux-based systems for its paint matching service in over 2300 stores. The benefit for Sherwin Williams is that they no longer need to pay expensive software licensing fees and, because they can also use free email and browsing tolls, they no longer need to pay for the privilege of using programs like Microsoft Outlook and Explorer. As an added bonus, because the software is open source their developers could customize the operating system to their own specific needs.

Linux has even found its way to Hollywood with animated hit movies like Shrek, which was produced using a cluster of Linux servers. The digital video recorder TIVO uses Linux to power its home entertainment units.

This all sounds too good to be true: Free, Customizable, Secure, Robust. So what's the catch?

For the longest time, Microsoft has pitched its strategy against Linux as they simply had better technology. Recently they have backed away from the direct technology head to head and moved towards more of a TCO (Total Cost of Ownership) argument. Meaning that while Linux is free, it is more expensive to maintain, support and administer on a day-to-day basis.

Linux lacks the management tools of those of the more mature Operating Systems like UNIX and the Microsoft Windows platform. A recent survey by IDC (commissioned incidentally by Microsoft) has noted that Windows may be a step or two ahead on such networking applications like file sharing and print sharing.

Linux still seems to be ahead of the TCO curve when it comes to applications like web serving. It is also important to look at the progress Linux has made in the last few years. What basically started out as an interesting academic project by a smart Finnish chap by the name of Linus Torvalds was transformed by the open source community and the new fangled world wide web into what is today one of the only software platforms in the world capable of unseating the mighty Microsoft! And IBM, Dell, Intel, Hewlett Packard and a whole host of other technology heavy hitters have bet heavily on the Linux horse coming home.

The next big frontiers for Linux to conquer are the consumer and corporate desktop market and correspondingly the network server market. Don't look for Microsoft to give up market share in these markets easily. Microsoft seems to have changed its message recently as it pertains to Linux, now opting for the "Windows is cheaper to maintain" argument rather than the "Windows is better technology" one, which they weren't really winning.

This new tack seems to be paying off in the short term as Linux's pace of adoption, will still is at an impressive 30 percent for 2003, seems to be slowing down. Of course, this could be more to do with natural adoption curves and the slowdown in technology growth due to events of the recent few years than any central Microsoft marketing campaign or "messaging strategy." It is likely that we will see many management and administration tools popping up for the Linux platform over the next few years. These tools will likely prompt organizations to use Linux servers for more than the traditional web serving applications. Linux is not the only open source

Linux 53

success - the open source Apache web server has over 62 percent market while its nearest competitor Microsoft is at 27 percent.

Linux will likely move from its current role as predominantly a web serving and parallel processing workhorse to playing a more central and traditional role in an organization's IT infrastructure. It will be interesting to see, as more and more of these Linux management tools appear, how tempted some of the software companies will be to make these tools proprietary. IBM's Tivoli product had only two management tools available for Linux just two years ago, now there are over 20.

It is also important to note that a large portion of companies already use what is essentially "open source" software in that many existing applications are "home grown" or written by internal developers for a specific purpose or task and the source can be viewed and tweaked by another developer, hence Open Source.

Of course the other advantage of Global open source software is that it is not only the development team, who are working under pressure, who get to see it and improve it.

Companies like IBM and HP have sunk huge amounts of dollars into Linux's Open Source research and development..

There are and will continue to be winners and losers in the Linux revolution. It seems that IBM and Dell are doing well from Linux by selling large amounts of mid and low end servers at the expense of people like Sun who generally make their nut from high end Unix-based servers. Companies like Etrade, the online trader, are replacing expensive large Sun servers with cheaper Linux-based boxes. This is also allowing Intel to piggyback on Linux into the corporate server market – a move that opens up an entire new market for Intel while at the same time decreases its interdependency on Microsoft. This goes a long way to explaining why Intel has been such a big supporter of the Linux movement. In the short term at least, both Sun and Unix have the most to lose from the Linux threat with Microsoft perhaps most vulnerable in the longer term. For example, the Chinese government has chosen Linux to power over 3200 post offices in one province alone.

Linux is also having a tremendous effect on computing in countries outside the western economies. In places like eastern Europe or South America, where expensive software licenses and high end hardware are hard to find, the free Linux and its attendant applications and cheaper hardware are making impressive inroads and allowing these economies to take advantage of the current age of technology we find ourselves in

So in the end there are a lot of players who are involved with and have a stake in Linux and the Linux community. From the head of IT at a large global financial company to the hordes of freelance developers who spend thousands upon thousands of their free hours debugging and improving to the administrator of a school district in Peru to the army of IBM programmers dedicated to Linux development to the Open Source revolutionaries and "technical freedom fighters," there are some strange bedfellows and the future of Linux, at least for now, largely depends on their ability to coexist and continue to push the Linux phenomenon forward. Time will tell.

While Linux has made amazing strides, the revenues from sales of Linux services are still a tiny fraction of the software market – a paltry 291 million in revenues of server operating systems by 2006 projected by IDC compared with a global market that was worth 10<6 billion in 2001. When we look at the numbers it is easy to think that everyone is overreacting to the Linux phenomenon.

Steve Ballmer, the CEO of Microsoft, has called Linux a "cancer" undermining the software model although he probably meant a cancer eating away at incumbent profits. We would expect him to say this. Imagine if some collective brought out a free car that worked as well if not better than all the other cars on the road and all you had to do was pay for the gas! One could imagine what the CEO of GM or Toyota might say!

It is also possible that over time Linux will end up with a subscription based model where customers receive the initial code for free but they pay for the updates and support. This is currently the model pursued by many of the Linux distributors like Red Hat, Caldera and SuSE.

So one might ask how is this different from the current license-based model that companies such as Microsoft prefer – well, in actual fact, it may not be that different in the long run – although one of the big differentiators currently is the fact that one can access the source code of the Linux OS which means it is highly customizable unlike say Solaris (Sun's Unix operating system or Microsoft Windows).

But all is not lost for Unix which still has a decent strangle hold on the upper tier of the computing market while Linux and Windows tend to compete at the mid and bottom tiers. Although the performance gap between Unix and Linux should close in the next few years and some may argue the gap is already closed, Linux has had difficulty competing in the desktop market due to such things as poor usability and lack of decent peripheral support. Many analysts believe that this could hurt Linux in the long run as corporations look for a soup to nuts integrated solution like Microsoft Windows. Others like me believe that Linux improves everyday and with the oncoming proliferation of ubiquitous Internet connected devices, Linux will play a huge role. It is projected that by 2006 PCs will not be the dominant device connected to the Internet - this is excellent news for the Linux supporters.

Linux is also closing the gap between desktops with the Linux offerings becoming more and more polished, although developing slick, easy-to-use user interfaces is not cheap and a price war with Microsoft is probably something that no company on the planet would relish. Microsoft has said that Linux is like receiving a free puppy: "it may be free but you still have to pay more to feed it and take care of it." And they may have a point but with the amount of change that this scrappy ragtag movement has wrought in just over ten short years, nobody is quite ready to write off Linux just yet. Many corporations are keeping a beady eye on this "puppy" with intense interest in what it may grow up into.

While Linux has definitely been a success story to date and may become an even greater one, there are still many questions and concerns that exist. The entire software business to date has been built on a simple model: build software, sell it, make a profit and continue to innovate and expand the capabilities on your software. Simple!

In the open source world where the software is free and each company developing it must agree on sharing their innovations and customizations of that software with everybody else, including competitors, the model as we know it seriously changes. One can see that this is not likely to be an easy transition.

This cooperation has occurred between large companies like IBM and Sun in the case of Linux development because of a large common enemy, Microsoft. But what happens when there is no common enemy? Will this cooperation last or will companies hoard their innovations for their own competitive gain (which, it is helpful to remember, is why most companies exist in the first place!)

Can Open Source flourish in the corporate environment? Can a model exist that will be both profitable and workable at the same time? Sure it can, but it won't be easy. The business model today for Linux (the de facto poster child for the Open Source movement) seems to be selling services and applications around the core free operating system. And that may not be a bad model in itself.

Look at how IBM has been slowly transitioning over the years from a hardware player to a service vendor. The entire model for their future seems to be built around the "On Demand" computing model where IT infrastructures and computing basically become a utility like phone, water or electricity where you purchase units from your local utility company. Just as most companies no longer have their own generators or telephone exchanges, it seems some day they will not host their own computers and databases. And if companies don't own hardware then it's likely that they will not own software, either; thus the entire concept of owning software licenses becomes rather mute. And the Linux movement better hope it does!

How will Linux Affect Businesses?

Linux has already had a huge effect on businesses looking for lower cost alternatives to Windows and Unix-based computer systems. According to a 2003 report by Goldman Sachs, 39 percent of US corporations now use some form of Linux. And that figure is likely to grow. Linux is also in the process of changing or at least affecting the very model on which the software development business is based. Migration costs are likely to be one of the main issues facing organizations considering the switch to Linux. Retraining staff is usually one of the biggest issues. It is worthy to note that organizations already using Unix normally have an easier time from a resource perspective than do those heavily invested in the Windows platform. If the Linux-like open source subscription model continues to gain popularity as opposed to the current licensing model, the organizations will be forced to rethink how they develop, deploy and procure software. And that will affect everyone. At the very least, Linux is here to stay and will continue to provide a computing alternative for businesses.

How will Linux Affect Consumers?

n the short term consumers will not directly feel the effects of the Linux shift. Linux will still remain in the realm of IT managers and tech hobbyists although that may change over time as Linux continues to make improvements in its desktop environment and usability. Don't be surprised to see a myriad Linux powered from microwaves to traffic lights in the near future.

What to Watch for in the Future:

Linux has done well providing a robust, less expensive alternative to Unix and Windows for certain server applications. It has done less well as a desktop replacement or an environment for large-scale initiatives such as databases or ERP (enterprise resource planning) systems for example.

Linux will continue to gain market share in the low-end server business and will proliferate in a whole host of embedded devices. The question for the Linux community will be can Linux continue making in roads into the enterprise while keeping the current loose coalition together that keeps each distribution common; in other words, can Linux continue to remain true to its Open Source roots? Another thing to watch for in the future of Linux is how much market share it can grab in the desktop market – this is seen by many analysts as key to its penetration of the corporate and consumer landscapes.

Most Frequently Asked Questions

What is an Operating System?

Operating systems are computer programs. An operating system is the first piece of software that the computer executes when you turn the machine on. The operating system loads itself into memory and begins managing the resources available on the computer. It then provides those resources to other applications that the user wants to execute. Typical services that an operating system provides include:

Is Linux more secure than Windows?

This is a highly contested question. While it is true to say that the majority of the large-scale security attacks to date have taken place on Windows, this may have more to do with the fact that Windows is inherently a much bigger target due to its large user base and mistrust of Microsoft in the hacker community than security vulnerabilities in the Windows OS. Linux may seem more secure on the surface due to the open nature of its distributed development and of course some may argue that this very openness may make it more insecure. If everyone knows exactly how it works then it is easier to unlock, right? Not necessarily but the one thing we do know is that to date Linux has not nearly been as challenged as Windows from an attack perspective. The irony here is that more mainstream Linux becomes, the more vulnerable it may become.

If Linux is free how come companies like Red Hat charge for it?

While the core operating system is free, the Linux software model works kind like a subscription service. Customers pay for support, service and updates of the core Linux distribution.

What's the difference between Linux and Unix?

Linux and Unix are both operating systems that stem from essentially the same roots. They have a lot of aspects in common which usually makes it easier for a Unix shop to switch to Linux than a Windows shop. However, they are different in the respect that Linux remains totally Open Source with one common agreed on distribution that everyone agrees on. Unix, on the other hand, has over thirty different varieties each with its own set compatibilities. And while there are some Open Source versions of UNIX, like FreeBSD, the majority of flavors are commercial ones offered by companies such as Sun, HP and IBM.

Is Linux ready for primetime?

Linux has already reached primetime with over 39 percent of US corporations using some form of Linux. The real question is how much further Linux will continue in its stellar ascent? My feeling is that while Linux will continue to grow its market share, its growth will slow and there will be challenges to its sole distribution base. However, as more and more devices gain processors and connect to the Internet, Linux has a true opportunity to dominate the non-PC based computing market. While Linux may not currently be ready to challenge Microsoft and Apple for the desktop market, it is making huge progress and this may be the next big frontier for the "little operating system that could" to conquer.

Buzz Words and Definitions:

Command Line- A space provided directly on the screen where users type specific commands. In Linux, you open a shell prompt and type commands at the command line, which generally displays a $ prompt at the end.

OS (Operating System)- The main control software of a computer system. The OS handles task scheduling, storage, and communication with

Linux

peripherals. All applications installed on a computer system must communicate with the operating system. Linux is one example of an operating system.

Source Code- Specially written instructions by a software programmer to create executable programs when run through a compiler or language interpreter.

Kernel- "Kernel" refers to the core OS. It is an abstraction layer between hardware and software, which provides an environment for running software programs. It contains large chucks of C and assembly codes for providing the functions of an OS. This is the part of Linux that is common for every distribution.

TCP/IP (transmission control protocol on top internet protocol) - Communications protocol used to connect to a variety of different types of hosts on both private networks and carrier networks such as the Internet.

UNIX- Unix is a multi-user, multi-tasking network operating system developed at Bell Labs in the early 1970s. Linux is based on, and is highly compatible with, Unix. Unix is not a single operating system. It is in fact a general name given to dozens of operating systems by different companies, organizations, or groups of individuals. These variants of Unix are referred to as "flavors." Although based on the same core set of Unix commands, different flavors can have their own unique commands and features, and may be designed to work with different types of hardware. Some popular "flavors" of UNIX are HP-UX, IRIX, Linux, NetBSD, OpenBSD, Solaris and Tru64.

URL (Uniform Resource Locator)- A publicly routable address for resources transmitted via the World Wide Web (WWW). URLs can be name-based (such as www.example.com) or address-based (such as 192.168.1.2).

MINIX- A small operating system that is very similar to UNIX that was written by Prof. Andrew S. Tanenbaum of Vrije Universiteit, Amsterdam, for educational purposes.

GNU- GNU is a recursive acronym for "GNUs Not Unix". It is a UNIX-compatible operating system using the Linux kernel developed by the Free Software Foundation. The design philosophy of GNU is to create a full-featured operating system composed of completely free software. Red Hat Linux combines several parts of GNU along with the Linux kernel.

Open Source Software (OSS)- Non-proprietary software in which the software source code is available and can be adapted by users to suit their needs.

Linux Distribution- Linux distributions are developed based on the Linux kernel, adding enhancements, packaged with software and tools for installation and configuration. Some of the more popular distributions are Debian, Red Hat, Mandrake and SuSE.

KDE (K Desktop Environment)- Graphical desktop interface designed with free software tools and libraries for the free software community.

GNOME (GNU Object Model Environment)- A graphical desktop environment for UNIX and Linux that is designed to provide an efficient and user-oriented environment.

The Bottom Line:

The bottom line is both simple and complex. Linux is here to stay and is probably already being used on some level in your company. The question at this point is not whether Linux will succeed but exactly how far it will go. Linux definitely has a place in the enterprise and can be very cost effective if deployed correctly but businesses need to be wary of some of the hidden costs like ongoing maintenance and training.

Top Ten Cheat Sheet: Linux

● Linux is an Open Source computer operating system developed in the early nineties and is based on UNIX

● Linus Torvalds is the Finnish student who first developed Linux and posted the code on the Internet for other programmers to tweak and improve, something they still do to this day

● Linux is essentially free

Linux

- Linux has been quickly adopted by IT organizations, as it is a robust, cheap and powerful alternative to UNIX and windows for web applications and cluster computing.

- Linux is beginning to make in roads into the desktop market with significant progress in embedded applications with everything from PDAs to TV set top boxes to cell phones

- IBM has invested more than a billion dollars in Linux and currently has more than 4600 Linux customers.

- Linux currently lacks the administration and management tools of the more mature Operating Systems like Windows NT and Solaris

- Microsoft has said that Linux is like receiving a free puppy, "it may be free but you still have to pay more to feed it and take care of it."

- Don't be surprised to see a myriad Linux powered from microwaves to traffic lights in the near future

- At the very least Linux is here to stay and will continue to provide a computing alternative for businesses.

6

Wireless Technologies – RFID, Bluetooth, Wi-Fi and GSM

"I find that the harder I work, the more luck I seem to have."

Thomas Jefferson

All of the technologies we will talk about in this chapter have two things in common – one, they all relate to wireless technologies and secondly, they all have the capability to change the way business is transacted.

By the end of this chapter you will be conversant with the fundamentals of RFID, Bluetooth, GSM and Wi-Fi and how these technologies will impact your business. While they have different names they are essentially radio based technologies and, at a simplistic level, could be said to differ because of one issue – their range.

RFID, Radio Frequency Identification, is a short-range system used to identify products in situ or in transit. RFID is used to track products on a shelf in a supermarket and can offer the retail industry a boon in saving on inventories and out of stock items. Already, Wal-Mart have told their major suppliers that they want to be fully RFID enabled by 2005.

Bluetooth is a clever radio system operating in an approximate 10 meter range that allows pairs of devices to communicate to each other, wirelessly. Toyota has developed their Prius car, a fully Bluetooth enabled automobile. Your cell phone will talk to the car's audio system and you can use buttons on the steering wheel to make and receive calls – all completely wireless. There's a good chance you already use a wireless mouse and wireless keypad in your office – the probability is that these devices are using Bluetooth to communicate.

Wi-Fi stands for Wireless Fidelity and allows for a wireless network to be established, with a range dependent on performance, which can reach hundreds of meters. Many offices are now using Wi-Fi as a cost effective alternative to running Ethernet cables all over their buildings. Starbucks now offer customers the opportunity to log on to their own Wi-Fi network – cappuccino and surfing in convivial surrounds.

GSM stands for Global System for Mobile communications and is a longer range radio based switching system used in cell phones. The system has migrated from its first generation, which was an analogue platform, to a fully digital platform for its second generation. Third generation or 3G handsets will be fully rolled out by networks in 2004. This will enable a full digital environment with the ability to send and receive email video images, internet connectivity and other enhanced services. When you consider that GSM is available in over 200 countries and will surpass the 1 billion customer mark in 2004, it is fair to say it is a truly global network.

Regardless of the range issues – one thing is certain – these radio technologies are an intrinsic part of our everyday lives.

Detailed Summary:

RFID

RFID has really been set alight by the endorsement of the technology by the retail industry. RFID tags are miniscule microchips, which already have shrunk to half the size of a grain of coffee. RFID tags work by listening for a brief radio signal and then respond with their own, completely unique ID code. The beauty of these devices is that they require no batteries – they are powered by the original radio signal. Obviously, they are disposal entities but when you consider that unit costs are in the 5 – 15 cent bracket you can see that their cost can be sunk – for the right benefits accruing, of course.

Many industry experts believe that RFID is ripe for widespread adoption. One driver of this is the Wal-Mart chain announcement that they want their top 100 suppliers to be RFID compliant by 2005. All suppliers will need to be in the loop by the end of 2006. This is a massive undertaking when

Wireless Technologies – RFID, Bluetooth, Wi-Fi and GSM

you consider the number of suppliers involved. Historically, companies, like Wal-Mart, needed a way of capturing accurate, real-time information about the products they make, move and sell. RFID offers that capability and helps companies boost supply chain efficiencies, reduce inventories, limit theft, improve product availability and add convenience for consumers.

Wal-Mart is not alone, thankfully. In late 2003, the Department of Defense's Logistics Agency set out a new policy that expands active RFID tracking to all military shipments including cargo, ammunition shipments, and supplies. Wal-Mart and DoF are leading the way but others will definitely follow. Tesco, the biggest retail chain in the UK have also adopted this technology by installing Smart Shelves with networked RFID readers. Consumer goods giant Gillette also announced that it would procure over half a billion RFID tags from a California based company. At these volumes the unit price of the tags, perceived by some as a barrier to early adoption, will come down.

The brains behind RFID were driven by the Auto-ID Center, based at MIT, an unusual cooperative effort between academia and global companies to develop the Electronic Product Code (EPC), a system for identifying objects and sharing information about them securely over the Internet.

One potential banana skin for the industry is the issue of privacy. It might be useful for you to consider this argument as it may arise within your own organization. This mainly centers around the size of the tags – they can fit seamlessly onto a garment or purse and will look like a speck of dust. The smallest one available at the time of print was a 550 microns square with a half inch antenna. This is small. Too small to notice as you leave a store, with the offending device attached to your clothing. This is not Mission Impossible technology – this is the here and now technology. Proponents of the technology insist that this is not a big brother technology – the range limitations ensure this – once a customer leaves a store the unique identifier code becomes useless. If the RFID tags are used on packaging then once the packaging is discarded there should be no problem. My advice is to watch closely how this technology unfolds and how the industry rolls out the PR of using a system that saves them millions of inventory wastage dollars. Part of the PR battle might be the carrot of reduced prices for the consumer.

Bluetooth

And now on to Bluetooth, a wireless technology that enables short-range wireless data connections between devices. One bit of trivia to impress your co-workers with is where the name came from: Harald Bluetooth, a Viking and king of Denmark from the years 940 to 981, was renown for his ability to help people communicate. During his reign, he united Denmark and Norway. My wife is Danish so I know exactly where the guy is coming from.

Bluetooth wireless technology is a worldwide specification for low-cost radio that provides links between mobile computers, mobile phones, other portable handheld devices, and connectivity to the Internet. There is a written specification developed, published and promoted by the Bluetooth Special Interest Group (SIG). This SIG includes Agere, Ericsson, IBM, Intel, Microsoft, Motorola, Nokia and Toshiba, and hundreds of Associate and Adopter member companies. In mid 2002, the Bluetooth SIG established its global headquarters in Overland Park, Kansas, USA.

The Bluetooth wireless technology is essentially designed to replace cables between cell phones, laptops, and other computing and communication devices within a 10-meter range. When Bluetooth wireless technology connects devices to each other, they become paired. An example of such device pairings includes:

- Your wireless headset connecting to the cell phone in your pocket

- Your PDA automatically synchronizes with your computer when you walk into the office.

And this is only the tip of the iceberg. Bluetooth Technology is poised to expand into areas such as industrial automation, gaming and delivery tracking. It is not too far off from when you will use a Bluetooth pen to write on an image board located in a different office. Your printer can be seamlessly and wirelessly connected to your PC. And anything else you can imagine in your office within a 30 feet range.

Bluetooth has already managed to immerse itself into the automotive industry. The new Toyota Prius is geared up to be Bluetooth enabled. Your mobile phone headset will be wirelessly connected to the car's in-built audio system enabling completely hands-free calls.

Wireless Technologies – RFID, Bluetooth, Wi-Fi and GSM

The Techie Bit

If you are interested in radio technology you may like to read this brief overview – if not, skip it – it will not affect your understanding of the fundamentals of what Bluetooth is all about and how it will affect your life and your business.

The Bluetooth wireless specification is one of the few wireless standards to include both the link and application layer definitions. Radios that comply with the Bluetooth wireless specification operate in the unlicensed, 2.4 GHz radio spectrum ensuring communication compatibility worldwide. These radios use a spread spectrum, frequency hopping, full-duplex signal at up to 1600 hops/sec.

The signal hops among 79 frequencies at 1 MHz intervals to give a high degree of interference immunity. Up to seven simultaneous connections can be established and maintained. This also gives Bluetooth devices a good deal of in-built security. Most of the time security will not be an issue due to range considerations but there is the flexibility to up the security should it be required.

The Good News

In late 2002, Microsoft announced that it would support Bluetooth in Windows XP. Apple Computer also announced that Bluetooth technology was supported in Mac OS X. Bluetooth was already supported in the Palm OS in late 2001 and Windows CE.Net in the first quarter of 2002. Support in these mainstream operating systems dramatically increases the opportunities for companies to integrate Bluetooth technology into new and existing products.

One excellent example of a large company utilizing this exciting technology is FedEx. They plan to integrate Bluetooth technology into their mobile fleets, allowing couriers to instantly retrieve and send shipping data to the FedEx operating system, without having to return to the van. FedEx is investing $150 million to roll out Bluetooth PDAs for its 40,000 US-based couriers, handling up to 50,000 packages per minute. With Bluetooth technology, they will link to the company's GPRS smart-phones, printers, and other devices

Wi-Fi

Wi-Fi, or Wireless Fidelity, is the next level up in radio-based technology from a range point of view. It allows you to connect to the Internet from your couch at home, a bed in a hotel room or a conference room at work without wires. Wi-Fi is a wireless technology like a cell phone. Wi-Fi enabled computers send and receive data indoors and out; anywhere within the range of a base station. And the best thing of all, it's fast. In fact, it's several times faster than the fastest cable modem connection.

However, you only have true freedom to be connected anywhere if your computer is configured with a Wi-Fi CERTIFIED radio (a PC Card or similar device). Wi-Fi certification means that you will be able to connect anywhere there are other Wi-Fi CERTIFIED products — whether you are at home, the office or corporate campus, or in airports, hotels, coffee shops and other public areas equipped with a Wi-Fi access available.

Wi-Fi Certification comes from the Wi-Fi Alliance, a nonprofit international trade organization that tests 802.11-based wireless equipment to make sure it meets the Wi-Fi standard and works with all other manufacturers' Wi-Fi equipment on the market. Thanks to the Wi-Fi Alliance, you don't have to read the fine print or study technical journals: if it says Wi-Fi, it will work. It's powerful.

Wi-Fi networks use radio technologies called IEEE 802.11b or 802.11a to provide secure, reliable, fast wireless connectivity. A Wi-Fi network can be used to connect computers to each other, to the Internet, and to wired networks (which use IEEE 802.3 or Ethernet). Wi-Fi networks operate in the unlicensed 2.4 and 5 GHz radio bands, with an 11 Mbps (802.11b) or 54 Mbps (802.11a) data rate or with products that contain both bands (dual band), so they can provide real-world performance similar to the basic 10BaseT wired Ethernet networks used in many offices.

GSM

GSM is a platform that allows cell phone users to roam within their own country or roam across international boundaries, keeping the same number and using a seamless billing system. There are over 200 countries now where GSM is active. Countries like Iraq are also opting for the GSM platform to enable them to roll out their cell phone systems.

Wireless Technologies – RFID, Bluetooth, Wi-Fi and GSM

The GSM system was originally developed for Europe and now claims to have over 71 percent of the world market. In fact, GSM orginally stood for Groupe Speciale Mobile from the original European working group that began the specification and standardization process. Initially the system operated in the 900MHz band but has been subsequently modified to operate in the 850, 1800 and 1900 MHz bands – giving it true global coverage.

	Example Application	Approximate Range Measurement
RFID	Inventory Tracking	Inches
Bluetooth	Wireless keyboard and mouse	Feet
Wi-Fi	Wireless Internet Access in office building, hotel, coffee shop	Yards
GSM	International cell phone roaming	Miles

So what are the Goals for these wireless technologies?

One of the primary goals for wireless technology is to ensure that adoption is not hampered by device availability. This was one of the problems with 3G – the technology was talked about so much before it was available yet the handsets remained discouragingly expensive. Vendors must ensure that wireless technology devices and networks are not priced out of the market.

Wi-Fi offers great potential cost savings when compared with wired office cable networks but it must be offered at a price that is still competitive. The GSM Association has stated that its primary goal is to continue the explosive growth of its technology platform adoption globally. Bluetooth has gone a long way in also ensuring that its technology is open – it can be used anywhere in the world between Bluetooth enabled devices. It will be interesting to see how RFID pans out – it has a slower acceptance curve because it is primarily focused on industry but widespread adoption of this technology will depend on consumers seeing the real benefits for them.

How will these technologies affect your business?

RFID- Retailers will generate significant savings in inventory and labor costs by adopting RFID technology.

Bluetooth- Bluetooth can significantly increase the mobility and the productivity of your workers in and out of the workplace.

Wi-Fi- Wi-Fi can afford your company the flexibility to operate in a completely wirefree environment. New staff can be seamlessly added to the network and be productive with zero delay.

GSM- Cell phone roaming is a boon for your company if you have international offices and need to contact staff all over the world.

What to Watch for in the Future:

The boundary between Bluetooth and Wi-Fi technology will become more seamless. Bluetooth technology and Wi-Fi are actually complementary technologies. Bluetooth is designed to replace cables between cell phones, laptops, and other computing and communication devices within a 10-meter range. Wi-Fi is wireless Ethernet providing an extension or replacement of wired networks for dozens of computing devices. Expect to see both Wi-Fi and Bluetooth wireless technology coexist. Your office will use wireless technology as a cable replacement for devices such as PDAs, cell phones, cameras, speakers, headsets and so on, whereas Wi-Fi will replace higher speed wireless Ethernet access.

With the advent of GRPRs and 3G you will see a lot more rich content being delivered with your cell phone. Surfing the internet at usable speeds and downloading email are just some of the features that will enable your cell phone to be an extension of your office environment.

Buzz Words and Definitions:

RFID- Systems that read or write data to RF tags that are present in a radio frequency field projected from RF reading/writing equipment.

Nominal Range- The range at which a systems can assure reliable operation, considering the normal variability of the environment in which it is used.

Passive Tags- Passive tags contain no internal power source. They are externally powered and typically derive their power from the carrier signal radiated from the scanner.

Tag- The transmitter/receiver pair or transceiver plus the information storage mechanism attached to the object is referred to as the tag, transponder, electronic label, code plate and various other terms.

802.11 WLAN - A Wireless LAN specification defined by the IEEE.

Bluetooth - An open specification for wireless communication of data and voice. It is based on a low-cost short-range radio link facilitating protected ad hoc connections for stationary and mobile communication environments.

IEEE - Institute of Electronic and Electrical Engineering

WLAN - Wireless Local Area Network

Wi-Fi- Wireless Fidelity, also known as 802.11b, a technology that uses radio waves for computers to connect to each other and to the Internet.

802.11b- Industry standard for wireless Internet use. Operates through radio frequencies around 2.4 GHz. Common electronics like cordless phones and microwaves also operate on this frequency.

Access point- Hardware that connects to existing DSL or cable modem line, essentially turning a wired connection into a wireless connection within a certain distance of the hardware. Access points are shared connections, so more than one person can access the Internet from them.

WEP- Wired equivalent privacy or wired encryption protocol, basic Wi-Fi security is used to protect data and Internet access from outside users. The encryption process uses algorithms to secure data being transferred via radio waves.

Router- A device that forwards data from one WLAN or wired local area network to another. The router is able to determine the fastest and most reliable way to send data from LAN to LAN.

GSM- Global System for Mobile communications, the second generation digital technology originally developed for Europe but which now has in excess of 71 per cent of the world market. Initially developed for operation in the 900MHz band and subsequently modified for the 850, 1800 and 1900MHz bands.

2G- The second generation of digital mobile phone technologies including GSM, CDMA IS-95 and D-AMPS IS-136.

2.5G- The enhancement of GSM which includes technologies such as GPRS.

3G- The third generation of mobile phone technologies covered by the ITU IMT-2000 family.

Bluetooth- A low power, short range wireless technology designed to provide a replacement for the serial cable. Operating in the 2.4GHz ISM band, Bluetooth can connect a wide range of personal, professional and domestic devices such as laptop computers and mobile phones together wirelessly.

Cell- The area covered by a cellular base station.

EDGE- Enhanced Data rates for GSM Evolution; EDGE uses a new modulation schema to enable theoretical data speeds of up to 384kbit/s within the existing GSM spectrum. An alternative upgrade path towards 3G services for operators, such as those in the USA, without access to new spectrum. Also known as Enhanced GPRS (E-GPRS).

GHz- Giga Hertz, a unit of frequency equal to one billion Hertz per second.

Wireless Technologies – RFID, Bluetooth, Wi-Fi and GSM

GPRS- General Packet Radio Service; GPRS represents the first implementation of packet switching within GSM, which is a circuit switched technology. GPRS offers theoretical data speeds of up to 115kbit/s.

Roaming- A service unique to GSM which enables a subscriber to make and receive calls when outside the service area of his home network, e.g., when traveling abroad.

The Bottom Line:

RFID is coming to a store near you – literally. Watch out for how the big retailers play out the PR battle with cost savings versus privacy.

Bluetooth will be part of your everyday work and social life – from the humble wireless mouse to sophisticated in-car audio and cell phone systems.

Wi-Fi will change the way work is enacted in an office. No need for cumbersome, expensive and inflexible cables.

GSM will enhance your ability to work away from the office, especially with the widespread adoption of GPRS and 3G.

Top Ten Cheat Sheet: Wireless Technologies

- RFID is a radio frequency, short range tagging and ID technology.

- RFID is a powerful technology that will change the way retailers think about inventory and stock-outs.

- Bluetooth is a radio technology that pairs devices so they communicate with each other wirelessly.

- Bluetooth operates in the global 2.4 GHz (Giga Hertz) band but uses frequency hopping and spread spectrum techniques to minimize interference and maximize security

- Wi-Fi is a wireless network solution with the power to completely replace office cabling systems.

- Wi-Fi can operate at speeds of 11 Mbps or 54 Mbps and is a number of times faster than the best cable or DSL access available, so it will not impair your network's internet access.

- Wi-Fi and Bluetooth are complementary – Bluetooth can replace the need for short cables (mouse, keyboard, printer, etc.) while Wi-Fi can replace the network infrastructure of the entire office.

- GSM is available to over 1 billion customers globally

- GSM will be superceded by newer versions of its platform when GPRS, EDGE and 3G come into play.

- All the above technologies are radio based and will radically affect the way we all live and work.

7

XML

"It is the mark of an educated mind to be able to entertain a thought without accepting it."

Aristotle

Colleagues and co-workers at meetings may often mention XML in hushed tones - your web designers will nod their heads knowingly. They will mention that your company needs to adopt a coherent XML approach for all its data handling needs, including delivery over the internet. Some managers will nod their heads in agreement, even though they have no idea what XML means to them or their company. Do not fret. By the end of this chapter you will be able to nod your head also, but nod it knowingly.

XML stands for *Extensible Markup Language.* XML could be viewed as the more interesting, more flexible cousin of HTML (Hypertext Markup Language), the language that designers use to build Web pages. In XML, designers are allowed to create their own customized tags, enabling the definition, transmission, validation, and interpretation of data between applications and between organizations.

A specification developed by the W3C, *World Wide Web Consortium*, XML is a pared-down version of SGML (Standard Generalized Markup Language), designed especially for Web documents.

XML is rapidly emerging as the *de facto* standard for passing data between applications through the web inside an organization and with other organizations.

XML is also one of the building blocks of Web Services (see Chapter 1: **Web Services**).

It is essentially aimed at improving the functionality of the web by providing more flexible and adaptable information identification. It is called extensible because it is not a fixed format like HTML (a single, predefined markup language).

XML has been heralded as the next important Internet technology, the next step following HTML, and the natural and worthy companion to the Java programming language itself. Thankfully, the major players – Sun, Microsoft and their ilk – have been intimately involved in the development (and hopefully the deployment) of XML. Even more thankfully, they do not own XML – is it owned by everyone and no one vendor can claim it as their real estate.

What you really need to understand is that XML is just another of what are called Standardized General Markup Languages (SGML) – basically languages that define the way data is presented. HTML, the ubiquitous web language is an SGML language. However, HTML is really just a clever presentation language – effectively only a formatting language. It is constrained by a rigid selection of tags that tell a browser how to render a piece of code on your computer screen. These tags are cast in stone and cannot be tampered with. XML lets you develop your own tags AND share them with others if you so wish. This is the beauty of XML.

Still confused about SGML and all these markup languages? Well, don't be. Even before HTML the airline industry had a markup language for its own use. Remember, SGML is just a way of configuring languages for a specific reason – and HTML and XML are subsets of this.

Although XML is a member of the SGML coterie, just like HTML, it offers a lot more. That's the good news. The downside is that is it not uniformly supported by the major browsers – as of yet. When this happens expect to be bombarded with all things XML and to get to talk about it a lot more with your web designers and systems integrators.

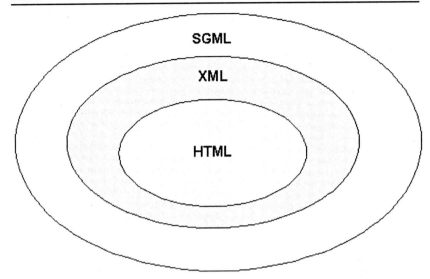

Figure 7.1 – Simple subset diagram for SGML, XML and HTML

Detailed Summary:

We have discussed XML is broad terms up until now. It may be useful for you at this stage to recap what XML actually is and what it is not. The table below should prove a handy reference for you.

XML is ...	XML is NOT ...
very easy to use and code. The complete specification document is less than 40 pages long. XML is designed to be easier to code than either HTML or full SGML.	a replacement for good old HTML. HTML is an excellent tool for displaying documents across a network. XML is designed for information providers who want to expand the horizons – something HTML is not geared up for.

thankfully, an open standard - XML is a subset of SGML (**see diagram 7.1 above**). This means that a lot of the hard work is already done for you. There are commercially available XML tools available on the market (XMLSPY from Altova is just one example).	a replacement for SGML. To make XML simple, many features of SGML were left out. People who are already using SGML may choose to use XML for network delivery, but there is no real need to convert existing SGML documents. XML is designed as an easy on-ramp to SGML for people who are not using it yet.
a learning curve - XML was designed by people with many years of experience, including members of the governing bodies for both SGML and HTML.	owned by any one company or organization. It is open for all to use. Nobody owns XML now, and nobody ever will, because SGML is a formally defined international standard.
by its very name - extensible - you have the power to invent and use your own tags and, if you choose, share them with others.	constrained to a defined set of tags.
very efficient - XML can re-use document elements and fragments, so you only have to transmit them once.	is not clunky – you can use it secure in the knowledge that it is as efficient as possible.
ready to use - web browsers can read XML today, just like HTML. You can use hyperlinks and images and multimedia, exactly as you do in HTML.	fully supported by all browsers – yet!
fully international - it has built-in support for texts in the major global alphabets, including a method to signal what language and encoding is being used.	

Hopefully the above table will put your mind at ease as to what XML is and what it is not. As you are, no doubt, comfortable and familiar with using

HTML (you use it every day you surf the net) it may be useful to look at a few of the problems that exist with HTML. This will help you to see why XML came into being in the first place.

A brief overview of some of HTML's Problems

HTML is an excellent presentation tool. It has enabled coders and non-coders alike to develop and publish web pages. However, HTML is easy for us non-coders because it is NOT actually a code. If you have used HTML you will see that the pages are defined using < > tags – and all these do is tell the browser whether to put something in bold, or in a paragraph or in a certain type of heading. You cannot, for example, tell the browser to do something with the piece of text depending on inputs from the user. It is merely a presentation and formatting tool. It is not a piece of software code like C++ or Java.

So, by that very definition, HTML is limited in what it can achieve. This is one of its problems. Web developers and systems integrators wanted to do more online but were constrained by HTML. Let us review a few of the issues that crop up when using HTML:

- HTML does not necessarily reveal anything about the information to which HTML tags are applied. For example, we know that <h2>Cocoon</h2> has a definite, predictable appearance in a web browser, but is it a XML-based web development framework? A movie with the much-maligned genius Steve Gutenberg? An ultra trendy bar in Dublin, Ireland? A home for a would-be beautiful butterfly? HTML does not usually tell. Not its fault – it just doesn't have the power to imbue this information. HTML tag names do not describe what content is. They only imply how content appears. This is a fundamental underperformance of HTML.

- As mentioned earlier, HTML has a fixed tag set. You cannot extend it to create new tags that are meaningful and useful to you and others. Only the W3C (World Wide Web Consortium) has the power to do this and they have better things to attend to (like XML).

- In the early days of the internet, web browsers were actually viewed as potential application platforms. However, technology like Java needs more power in the engine than HTML can offer to deliver on that vision. With HTML as the data standard, web-based applications rely heavily on CGI scripts, at the server, to process the data in web pages. This has the downside of contributing to internet traffic and makes the web slow for many users.

The above points are not intended to denigrate HTML, but merely to establish the perspective that some of XML's developers hold. HTML has served the team extremely well, but like every good quarterback knows, there is always some kid on the sidelines waiting for their shot at a call up to the major league.

So what are XML's Goals for the Internet?

As internet usage soared, people did more and more clever things with the existing HTML they had at their disposal. Some developers, implementers, and users of HTML's parent technology – SGML – looked on knowing that there was a more powerful entity that could be developed. These individuals and their companies had already invested heavily in SGML. They knew that with SGML they could govern the structures and describe the contents of their documents and the information of which the documents were composed.

SGML, unlike HTML, provides its users with an extensible tag set, and it establishes the rules by which documents are created. SGML yields sets of tags, as HTML is a set of tags, for characterizing what pieces of information are. SGML experts and structured information systems people believed that SGML technology could enrich and revolutionize the web in some key ways:

- **Electronic Data Interchange support**

    One of the principal uses of structured information is to enable electronic data interchange (EDI). Different industries create consortia to specify the content model on which they all agree, and which they use to mark up their information, so that they can share it with each other easily and efficiently. In the jargon of structured information, that content model is a DTD (Document Type

Definition) or a schema. Based on this premise it is prudent to think of the web as an ideal venue for electronic data interchange. Fledgling XML developers could envision a range of EDI applications for which HTML was an inadequate data format, ill-equipped to express an industry's content model and its rules of logic.

- **Java technology - and client-based processing**

    We talked earlier about how HTML could not open up the true power of technologies like Java. XML technology enables browsers to function as generalized application platforms. True platform independence is the result. But the fixed tag set and presentational limitations of HTML provides little for Java applications to process. To coin a phrase by one XML developer, "XML gives Java something to do."

    By providing information rich in metadata specified in a standard format, XML and Java technology make it possible for more of an application's work to be processed at a client. This contrasts with the general tendency of HTML pages to rely on a CGI script back at the web server for any programmed functionality, with all the incumbent web traffic and server loading issues I alluded to earlier. With XML and Java technology, more client-based application processing could reduce network and internet traffic, making the web faster.

- **Platform-independent information**

    SGML, the parent technology of HTML and XML, has always offered itself as a platform-independent technology for specifying the structure and describing the content of documents. While enterprises wrestled with evolving information formats like Microsoft's RTF, Adobe's PostScript and MIF formats, formats from WordPerfect, IBM, and so on, SGML represented a rigorously consistent and platform-independent form for representing information.

    However, during the 1980s, when the SGML standard was quietly emerging, most computer industry observers focused instead on the explosion and excitement of new computer platforms. That industrial and commercial momentum clouded the coming chaos that multiple proprietary information formats assured. Later, in the 1990s, the popular discovery of the internet, and the emergence of the web, web browsers, and Java, revealed that chaos more clearly.

As you can appreciate from above, XML has some lofty goals. It will take widespread adoption of this technology before these aspirations come to fruition. But XML can achieve this because it is not, in fact, some obscure coding language. Remember, XML is actually a "meta language" - a language for describing other languages - which lets you design your own customized markup languages for limitless different types of documents. XML can only do this because it's written in SGML, the international standard meta language for text markup systems (officially known as ISO 8879).

Let's look at a simple example.

To put the whole SGML and XML argument in some perspective, let's look at a real life example of how one body of people have developed their own meta language. Hopefully this will help you understand where XML fits into the SGML scheme of things.

For years the problem of encoding mathematics for computer processing or electronic communication has existed. The common practice among scientists before the web was to write papers in some encoded form based on the ASCII character set, and email them to each other. Several markup methods for mathematics, in particular T_EX, were already in wide use in 1992. The web revolutionized the way people made information available to each other. Witness the vast number of private individuals from the scientific community who put up excellent websites to share their findings.

However, even though the World Wide Web was initially conceived and implemented by scientists for scientists, the possibilities for including mathematical expressions in HTML has been very limited. At present, most mathematics on the Web consists of text with images of scientific notation (in GIF or JPEG format), which are difficult to read and to author, or of entire documents in PDF form.

I do not want to send you into cold seats remembering high school math but consider the equation $2^{2^x} = 10$. This image (or equation) is designed to fit the surrounding line in 14pt type. Of course, for other font sizes, the equation is too small or too large. Also, the original image for the equation was generated for a white background and if a user has their browser set to a different background color then major problems will ensue. Moreover, center alignment of images is handled in slightly different ways by different browsers, making it impossible to guarantee proper alignment for different clients. Thus ends the math lesson but what I wanted to demonstrate is that there is an SGML solution to these representational problems.

Thankfully, for all scientists and mathematicians there is now MathML, a specialized markup language which uses XML as its base syntax. MathML allows complex math formulae to be presented and transported across the internet thanks to its being based in XML (and thus being part of the SGML group). This means that there is an open standard for the proper online manipulation of formulae and equations. MathML is fully underscored by the World Wide Web Consortium (W3C) and should be supported by most of the major browsers. This would not have been possible with standard HTML. This is all the math I will allude to in this chapter so you can relax now.

How will XML Affect Businesses?

One thing to keep in mind about XML is that, no matter what part of the information process you are involved in, XML will play a part – whether with or without your knowledge of it. The clever use of mechanically interchangeable parts is what helped fuel the Industrial Revolution. The clever use of reusable electronic information is what fuels the Information Age.

The world of computing architecture is in a state of flux. Not only are the computers themselves changing, but new network computing approaches are continuing the evolution of the web, grids and peer-to-peer networks. These ever more distributed processes are flagging questions and providing new opportunities to help us manage information glut.

Businesses now have an increased need to interact with data. Vast numbers of people now need in-depth knowledge of information and of how to build systems to manage it. XML is here to help them achieve this but it is only when it is universally accepted that the real benefits will accrue.

What to Watch for in the Future:

Once you have created a body of XML information, you will learn to treat it differently from the information you had before. The applications, file systems, and other software you relied on to elaborate information may not work so well with XML. Those traditional tools may not effectively expose the new value in your XML information. But again there is good news. It is clear that the marketplace is well prepared to deliver XML support in all phases of an enterprise's transition to XML. Already many software vendors are announcing, testing, and even delivering tools to aid in these critical phases of your transition:

- Converting your legacy information into XML and structured formats that reflect your information's full value

- Developing new, XML-structured information, and evolving your newly converted information

- Managing the base of XML-structured information that results from your transition

Estimated Timeline:

As with most new technologies, there is no one date where they finally arrive and are completely adopted. One point where we can start is to say that the XML 1.0 recommendation was adopted by the W3C in February 1998 – and is thus only six years old, at the time of printing. This is still very young and thus we need to give XML more time to grow and mature. The first version of MathML, used in the example earlier in this chapter, was developed in July 1999, very shortly after XML was formally adopted.

What is of primary significance is how the major browsers roll out their support for XML. They already offer a huge degree of support but they need to undertake more to fully support XML in all future versions of their browser products. Once this occurs you can guarantee that XML will be everywhere.

XML

Buzz Words and Definitions:

XML is peppered with many acronyms. You would do well to become familiar with the ones covered below:

SGML (Standard Generalized Markup Language)- The standard markup language used by the publishing industry to specify the format and layout of both paper and electronic documents. SGML is very flexible and feature-rich, and it is very difficult to write a full-featured SGML parser. As a result, newer markup languages requiring fewer features (e.g., HTML and XML) are subsets of SGML. SGML is defined by the international standard ISO 8879.

What is XML? XML stands for Extensible Markup Language. XML is a system for defining, validating, and sharing document formats. XML uses **tags** (for example emphasis for *emphasis*), to distinguish document structures, and **attributes** (for example, in <A HREF="http://www.aspatore.com/">, HREF is the attribute name, and http://www.xml.com/ is the attribute value) to encode extra document information. XML will look very familiar to those who know about SGML and HTML.

Is XML SGML? Yes. XML has been carefully designed with the goal that every valid XML document should also be an SGML document. There are some areas of difference between XML and SGML, but these are minor, should not cause practical problems, and will almost be certainly reconciled with SGML in the near future.

Is XML HTML? No. An XML processor can read clean, valid, HTML, and with a few small changes an HTML browser like Netscape Navigator or Microsoft Internet Explorer would be able to read XML. The biggest difference between XML and HTML is that in XML, you can define your own tags for your own purposes, and if you want, share those tags with other users.

If the browsers remain tied to the fixed set of HTML tags, then XML will simply be an easy on-ramp to SGML, important probably more because the spec is short and simple than because of its technical characteristics. This will probably still generate an increase in market size, but not at the insane-seeming rate that would result from the browsers' adoption of XML.

The Bottom Line:

Whether you like it or not, XML will be pervasive and completely embedded in the systems you use. If you think HTML is everywhere on the web think of XML as HTML's bigger, bolder and more brazen brother – it will be all over the web but also all over your databases, systems and networks.

Top Ten Cheat Sheet: XML

- XML will be everywhere before you know it
- XML is a meta language
- XML is part of the SGML group of languages
- XML is at the heart of the web
- The major browsers have pledged support for XML
- XML is not constrained to a fixed set of tags
- HTML IS constrained to a fixed set of tags
- No one organization or company owns XML
- XML is open for all to use and develop
- XML will be part of all information interchange

8

Customer Relationship Management (CRM)

"A friendship founded on business is better than a business founded on friendship."

- John D. Rockefeller (1874-1960)

Even though this book is entitled 10 Technologies Every Executive should know, it would be folly of us not to cover Customer Relationship Management (CRM). In essence, CRM is not a technology, it is a philosophy. However, it is a philosophy predicated firmly on technology. Whether it is a basic relationship management tool or a sophisticated enterprise solution, CRM is a business imperative rather than something you simply need to know about.

The real challenge facing you today is how you leverage all this new technology to create and promote better customer relationships that drive loyalty and reduce churn.

There are many definitions of CRM but the one I think encapsulates it best is the following – CRM is the ability to recognize customer interactions with your business and manage, analyze, and optimize those interactions.

Customers have gotten used to what could be called the "Amazonification" of services. While many established companies were busy erecting brochure-ware websites, companies like Amazon.com realized from the beginning that they were in a commodity business and that true customer service and experience would be the differentiator. So they spent billions of dollars of venture capital on researching and improving their customer service personalization and CRM tools to the point where they are almost the gold standard for online service and interactions with their customer base.

While your company may not sell books or music on the web, there is a good chance your customers do buy books and music on the web – and therefore a good chance they have been exposed to such a company as Amazon.com.

The problem here is that customers have gotten used to this level of service and have come to expect it from your company and any other company they interact with online. This is the challenge facing you and your company. CRM is something you have no choice about – it is an imperative.

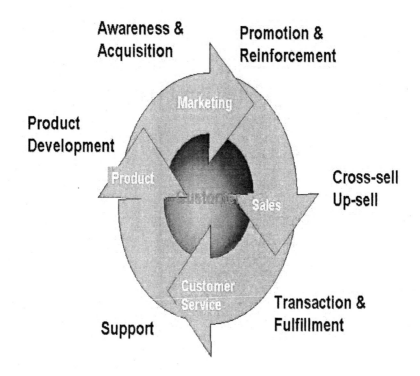

Customer Relationship Management (CRM)

Figure 8.1 – CRM overview

Detailed Summary:

You already engage in CRM. This is a fact. If you are in business you have to be engaging customers. This engagement may be as simplistic as your staff answering the phone right up to deep account tracking and using supply chain management software. The fact that you already engage in CRM does not necessarily mean that your organization is doing it correctly. Your company may run (or outsource) a call center and feel that this adequately caters for all your customer service interactions.

Your company may also have an expansive website that is used to cater to the needs of your "online" customers. Are the two channels mutually inclusive? True CRM should cater to all of your customers' interaction requirements in a seamless manner. The customer should not have to bother about what channel they use to contact you. That is your problem.

There is no way around it – to implement CRM properly you need to have the right technology. Rather than trying to bamboozle you with the technology requirements there is one thing this chapter should reassure you of – implementing CRM need not be a hugely expensive project. Whilst some CRM applications are extremely complex, especially supply chain management technologies, customer relationship management need not be. Much of the functionality offered through customer relationship management technology is quite straightforward.

In fact, customer relationship management is amenable to being offered as a hosted application. Salesforce.com is an example of one company who has geared their products to be offered as a hosted (off premises) solution.

Whether you opt for a hosted or enterprise solution I believe it is crucially important for you to realize that, in our world of increasingly commoditizing products and services, it's the customer experience that matters most. There are three basic tenets of good CRM that are worth stating here:

- Good customer experience will drive customer loyalty.

- Good customer loyalty will drive customer purchases.

- Good customer purchases will generate revenue.

No matter what business you are in, the above tenets are applicable to you. They are immutable laws that must be adhered to if you want to get the CRM problem cracked.

It is probable that the vast majority of your customers are using both traditional and online customer interaction channels. Whatever channel they choose – they expect one thing – not to have to repeat themselves once they change over from one channel to the other. They expect a fully seamless integration of all your systems. It is immaterial to them that you have trundling legacy systems that are hugely expensive to integrate. They may also be unaware that your organization has no plans to do this in the near future.

This is really the nub of what CRM is all about. The integration of your organization's service processes is by far the biggest problem that you will face. Spending and investing in efficiencies of individual areas of your business will count for nothing if you get the CRM element wrong.

Consider the experience of a prospective customer seeking a loan from a major bank, with extensive online and offline channels. After being referred to the website, the customer patiently completes an online form and submits the requested information, only to be told that he has made a mistake. With some irritation and frustration the online customer (or potential customer) tries again and is then told, after downloading five different screens, that he des not qualify for this particular type of loan.

Growing frustrated, he picks up the phone and repeats his story to a customer service representative (CSR) from the beginning, because the service rep has no record of his previous online interaction. It might turn out that the rep is able to tell the customer that he is, in fact, eligible for the loan but that the way the information was requested online precluded him from entering some key credit information. This scenario may solve the short-term problem for the customer – i.e. getting loan approval – but where will he go, and which channel will he use, when he next has a requirement to deal with this financial organization.

Does this example sound familiar to you? It is most probable that you have experienced this yourself (hopefully you got approval for that loan!). Recent

Customer Relationship Management (CRM)

industry studies have found about one-third of customers have had this experience when following up with a CSR after online service requests. It is simply not good enough.

Think of another side to this story – possibly one from your competitor. The person filling out the online loan application comes across the same problem. Again he tries it once again only to be told that he is not eligible for the loan. This time when he calls, frustrated, the CSR has all the details inputted online, available for discussion.

Think how impressed that customer will be, knowing that he does not have to divulge all of the personal financial details to the CSR again. He may still experience disappointment, perhaps not getting loan approval, but at least he has encountered a positive customer experience.

There is no question, even in a case where a customer does not achieve their ultimate objective (in the above example securing a loan), that they will be more impressed and hence more loyal to the company that offered the best customer experience. On the other hand, if your customers become frustrated, ignored, and confused, then you run the risk of losing their business. A recent Genesys survey showed that 85 percent said that a negative call center experience could stop them from doing business with a company and 56 percent said that they already had stopped doing business with a company after a less than satisfactory contact center experience.

Regardless of what surveys may highlight, one fact is agreed upon by nearly all customers - it is important to share customer service information across all interaction channels, whether it is web, phone or email. Forrester Research reported a number of years ago that fewer than two percent of companies have the ability to take a unified view of customers across sales, marketing and service departments. This is improving but the pace of improvement needs to be increased.

So what about investing in CRM?

Corporations agree that customer service is important to their livelihood. However, the majority of companies still consider their customer service departments as cost centers. When business is under pressure it may well be that these cost centers are the first to be hit - short-term cost cutting in hard times. Unfortunately, these quick fixes can obscure the need for a longer-term strategy that turns a company more fully toward the customer. There is a growing body of research that shows the customer experience is composed of not only what you provide, but also how you provide it.

One of the first hurdles in implementing CRM in an organization is executive sponsorship. In large corporations, executives' responsibility is spread across many business units, making it difficult to get consensus on a CRM strategy. Lack of consensus at the highest levels of an organization leads to more CRM project failures than any other element.

SMBs may not encounter the same consensus issues – perhaps the company owner or CEO makes all the investment decisions. The added costs should be a little more palatable given the jumps in functionality and new choices in delivery modes that SMB CRM vendors are offering, for example, NetSuite's product offering includes a direct link to UPS shipping, which may benefit some companies.

A large corporation, like a cable company for example, may have disparate brands – each with their own marketing department. Each of these divisions must be in agreement with the goals of the CRM project for it to be rolled out successfully across the organization. Getting consensus across a large corporation like this may be very problematic – and this is at the project specification stage – just imagine what might happen once it is required to roll out software to enable the project. The corporation's individual marketing departments may rely heavily on the services of third-party providers like advertising and marketing agencies. It is important to get buy-in from these business partners and to assuage fears they may have about potential loss of business.

A Brief Overview of Some of the Problems:

One problem that may be encountered when a CRM implementation project is at an advanced stage is the whole area of training. If a project is running behind it is possible that a company might cut back on the training. Remember a CRM solution is only as good as the links connecting it. When all the various customer interaction channels are integrated seamlessly it is imperative that the human factor is catered for.

Customer Relationship Management (CRM)

Small and mid-size businesses do not have to feel left out of the benefits that CRM can offer. Indeed, SMB have a lot more benefits to reap. CRM product offerings for SMBs are more cost effective than enterprise solutions and can offer a greater return on investment. SMBs can now choose from a wide range of CRM products – both hosted and on-premises. These companies include Microsoft, Salesnet, Netsuite, SalesLogix, Accpac, Onyx and Pivotal.

At the time of writing an SMB might spend as little as US$2,000 a year on a CRM solution, hardware included. Some mid-sized companies may end up paying up to the order of $4,000 but these figures still compare well stacked up against a standard implementation of Siebel, which can cost up to $12,000 per user.

One of the major growth areas over recent years has been in what Application Service Providers (ASP) can offer. Simply stated, an ASP is a service provider whose specialization is the implementation and ongoing operations management of one or more networked applications on behalf of its customer. The emphasis is on web-based e-business application management as opposed to the more traditional outsourced client-server application management services. ASP has also been known as Managed Application Provider (MAP) but the ASP acronym is more used now.

ASP CRM vendors are projected to expand their market share significantly. Siebel, for example, is jumping into the ASP space again with its OnDemand application, at the same time offering a migratory path to its premise-based application. This is in essence a hybrid offering. Companies can jump on-board and tip their toes in the water with the OnDemand hosted solution – and eventually migrate to the on-site version of their product.

PeopleSoft also announced an expansion of its hosted offering service to reach more customers. Most of PeopleSoft's enterprise product lines are now are available on a hosted basis. The fact that most of the big players are either entering (or returning to) the hosted product offering says a lot about how this market is evolving. The commitment certainly appears to be there from the CRM vendors.

Is CRM applicable to your sales team?

One question that often surfaces about CRM applications is how beneficial (and practical) are they for a team of sales reps out on the road. Supporting mobile CRM application users on the road has never been easy. Your team of sales reps of technical support staff may need to access different types of information while on the road. A sales rep might need to access latest, up to the minute pricing information or check on inventory levels before committing to a major sale. Technical staff on a customer site might need to access historical customer details or access particular product specification information. Many will use a cell phone and a laptop computer – and in a lot of cases this will suffice. Some may even use the information stored on their laptop, which at worst is a few days old, but is adequate for a particular requirement.

You will find, however, that some of your team may have a requirement for up to the minute information. I mentioned the inventory example above – this could be crucial for large sales of time sensitive stock. Thankfully there is a way to attend to the needs of these people. In an earlier chapter we discussed Wi-Fi (Wireless Fidelity) – which effectively gives broadband access while on the road, replacing GPRS or cell phone laptop configurations. All that is required is that you are within the coverage range of a Wi-Fi network, normally a few hundred feet or so from the access point. Normally, a nominal charge is applied to use the network.

Hundreds of wireless "hotspots" are popping up all over the globe. Starbucks now offer the cappuccino and surfing option. Once you are in one of their stores you can access broadband wireless internet access. This is a boon for on-the-road staff. Many hotels now offer this Wi-Fi service as well so there is no excuse for the traveling members of staff not to stay in touch. Before a crucial meeting with a client your sales rep could be coffee'd up and up to speed with the very latest inventory information. This is a very powerful tool for the on-the-road worker.

It should be noted that Wi-Fi networks are mainly concentrated in metropolitan areas. But with many hotels, motels and gas stations getting in on the play there should always be a wireless access point within a reasonable distance.

Customer Relationship Management (CRM)

One good example of who might benefit from using a mobile CRM application is the pharmaceutical industry. Picture a rep out on the road meeting doctors and distributing drugs to them. They may need to access historical data relating to shipments and usage of a certain drug. Under the PDMA (Physician's Drug Marketing Act), **they** may also need to capture a signature and track the drugs given to the relevant doctor. This could **tie** up very neatly with another technology we discussed in the book. If the drug samples (or product) are given to the doctor using RFID (Radio Frequency Identification) tags and these are then **zapped to base** using a Wi-Fi link, it would have a huge impact on efficiencies. The mobile CRM application would be updated in real-time and the rep could move on to the next appointment.

So what are the Goals for your company to drive CRM?

If you decide that CRM, as currently practiced by your organization, is not adequate or that you want to invest in a completely new CRM system, you may want to consider the following goals:

- Define a vision for the customer experience that will lead to loyal customers.

- Develop a compelling business case, with measurable goals and benchmarks, to help your organization make the right investment decision and track progress against a baseline.

- Use a cross-functional team and implement with a phased approach. Get short-term benefits while working towards your long-term vision.

- Get user feedback and make mid-course corrections to your multi-channel program as needed.

- Keep abreast of developments in technology and do not be afraid to integrate these into the project.

What to Watch for in the Future:

We have discussed the various channels open to the customer to contact your company – web, email and phone. Progressive companies are recognizing that instead of a call center to cater to customer interactions, they are developing e-service centers with full channel integration. A very small percentage of companies currently offer this level of sophistication. To develop such an integrated system across all customer channels for a large corporation may take up to five years to achieve.

What you will see is more and more companies start out on this journey and the ones that take full e-service CRM seriously will be at a distinct advantage over those that do not -- remember the loan approval service we cited as an example earlier.

A lot of companies are investing in updating their ageing telephony systems – introducing routing, queuing, voice-over IP and voice recognition functionality. This is all very well but the opportunity exists to invest in full e-service channels and not just a world-class telephony customer service. It is the potential integration between the e-service platform and the business application that creates true value for both the caller looking for a fast, effective resolution to their problem, and the business manager who is looking to increase the volume of callers.

The real challenge is to ensure that a customer contacting your company, by whatever means, is immediately identified and directed to the appropriate customer service rep. The agent will be presented with all of the appropriate customer details and any historical data relating to their issue and current status. One recent report from Phillips Infotech predicts that 46 percent of call center agent seats will be based on multi-platform technology by the end of 2003. Companies are recognizing that investment in e-channels of customer interaction is the way to improve loyalty and retention.

Other technology advances to watch out for in the CRM area is the use of voice recognition. A rep can access customer information from a database using voice commands. This is still in a relatively early stage of development but using the current systems a sales rep could read customer emails, check on meeting appointments and update client information. Surebridge offers a voice recognition system of Microsoft's CRM product.

Customer Relationship Management (CRM)

Sales reps can also use their Blackberry, using wireless technology to access CRM applications. In this scenario, the mobile worker types a text question addressed to the CRM application into a RIM Blackberry. The question is translated into a CRM database query and the answer is wirelessly e-mailed back to the user.

Voice and Blackberry access are only a couple of examples of how technology will assist with CRM application access. Technology is being developed that will make life for mobile CRM users much easier. This will include device specific improvements – longer battery life, better connections in "noisy" environments, more user friendly screens and interfaces, the integration of RFID reading systems in hand-held devices – the list goes on. The basic premise is that vendors are taking CRM and mobile CRM applications seriously and are rolling out technology to make their adoption easier.

Buzz Words and Definitions:

Application Service Providers- Application service providers offer an outsourcing mechanism whereby they develop, supply and manage application software and hardware for their customers, thus freeing up customers' internal IT resources.

Business Intelligence- Business intelligence refers to the type of granular information that line managers seek as they analyze sales trends, customer buying habits and other key performance metrics of an organization.

Business Metrics- Business Metrics is a set of traditional and nontraditional business measurements - such as judging product and service quality, rating customer relationships and measuring employee satisfaction and commitment - that are seen as critical for improving a company's bottom line.

CRM Applications- CRM applications can include capabilities such as sales force automation, employee help desk, customer support, and relationship enhancement. Usually, CRM applications implement CTI capabilities, giving them the ability to "pop" the appropriate screen, based on the customer's phone number (in the case of a telephone call) or the information collected from an IVR.

Computer Telephony Integration (CTI)- Computer Telephony Integration (CTI) is an architecture that allows telephone applications and services to be merged with computer applications and offered on open, accessible platforms. CTI simply coordinates the actions of telecommunications and computer systems to link the real-time call control capability of the telephone with the stored intelligence of computers.

Customer Retention - The means of satisfying a customer through a high quality level delivery of service and retaining their loyalty to purchase future products and services.

Customer Acquisition Costs- Customer acquisition costs are the marketing and advertising expenses needed to turn a prospective customer into an actual customer. This can include money spent to entice a visitor to a website or to a bricks and mortar store to purchase goods or services.

eCRM- Is the acronym for electronic Customer Relationship Management. This is the online version of Customer Relationship Management utilizing and interfacing business processes and data with offline, back end systems.

Inbound & Outbound CRM- Inbound CRM covers the experience customers have when they initiate contact with a business through any channel – whether through a call-center, IVR or the Internet. Outbound CRM is when a business initiates contact with a customer for the purposes of maintaining its relationship with that customer.

IP Contact Center- IP stands for Internet Protocol. An IP Contact Center is a multi channel contact center, designed to support customers using IP based telephony. Traditional specialized hardware such as ACD, PBX and IVR are replaced by applications on the IP Network. This results in a framework that is more flexible and provides a better platform for unified queuing and incorporating IP-based new media.

Voice Over IP (VOIP)- Voice Over IP, a term that originally described the transmission of real-time voice calls over a data network that uses IP, but currently is used to describe "anything over IP," for example, voice, fax, video, etc.

Customer Relationship Management (CRM)

The Bottom Line:

The bottom line is simple. Implement appropriate CRM for your organization or else sit back and watch customers migrate to your competitors. This chapter has shown that CRM is challenging to get it right but can reap major benefits for your organization once incorporated in all of your customer contact channels.

Top Ten Cheat Sheet: CRM

- CRM is a philosophy not a technology
- CRM can be empowered through technology
- CRM technology is only as good as the people using it
- For smaller companies it may be worth adopting a hosted CRM solution
- Larger corporations may require enterprise level solutions tailored to their requirements
- eCRM is the way of the future
- Getting your companies CRM implementation right will reap benefits
- CRM will generate loyal customers
- CRM will generate revenue
- CRM needs buy-in from executive management to ensure it is implemented properly

9

J2EE

"Do, or do not. There is no 'try.' "

- Yoda (*The Empire Strikes Back*)

Before we can have any discussion about J2EE, we must first mention the Java programming language, on which J2EE is based.

Java began life in the early Nineties as a platform-independent programming language created by a small group of engineers at Sun Microsystems. Platform independence was up until that point unheard of as all programming languages were linked to specific CPUs and Operating systems. The tagline for Java was "write once, run anywhere." Java may have remained just another interesting blip on the technology landscape where it not for the emergence of the World Wide Web. The Web and the browser were the perfect allies to promote the capabilities of Java and its initial "applets" that allowed browsers to do things with pictures and sound that simple HTML couldn't. Java took off like a rocket.

Since then Java has become a de facto standard for Internet and enterprise computing. The Java 2 Enterprise Edition (J2EE) is a set of standards based on porting the Java programming language to multi-tiered enterprise applications. You cannot download J2EE; it's not a software package. It's a set of rules that allow different software packages to work together. One of the main advantages of Java is that it is platform independent – meaning it will run on almost any operating system so developers can "write once, deploy anywhere." J2EE basically takes this portability one step further and applies it to multi-tiered enterprise software. In theory, with J2EE, organizations can employ a selection of best of breed software packages and as long as they are all J2EE compliant (that is they have been built with the same set of standards), they do not have to worry as much about the system integration of yesteryear. This enables them to be less dependant on one particular vendor. J2EE is the result of a large multi-vendor initiate led by Sun Microsystems, the original creator of Java. Other members of the initiate include IBM and BEA systems.

Detailed Summary:

J2EE is ultimately about choice. By settling on a standard and not on a particular vendor, organizations can have many more choices in how they approach existing system integrations and new technology initiatives. By limiting itself to one vendor a company may has less relationships to manage but they may be severely limited in the options and leverage they have. Before J2EE, to adopt a best of breed approach meant costly and complex integrations. J2EE, while by no means does away with system integrations, does however make those integrations a lot simpler and less time consuming and cheaper.

There was a point in time, and we're not fully passed it yet, where many executives would feel their temporal lobes shutting down at the very mention of the words "system integration" by the tech folks. Visions of millions of dollars and armies of consultants swam by their minds, Images of horribly complicated plumbing diagrams were conjured. At its very essence that's what system integration is, plumbing. The interconnection of several disparate computing systems and programs to accomplish shared tasks and pass data back and forth.

Connecting your bathroom sink to the water system for input and the drainage system for output is a form of system integration. Fortunately in the "real" plumbing world there is general agreement on the types of pipes and connectors to use to make this system work. This hasn't always been the way in the "electronic" plumbing world. Envision many different sinks and faucets and drainage systems and each one with a different size connector!

No wonder the nightmarish visions in certain quarters when it comes to system integration.

J2EE is an attempt to make all those "electronic" connectors the same size, i.e., talk the same language so they can connect to each other without much difficulty.

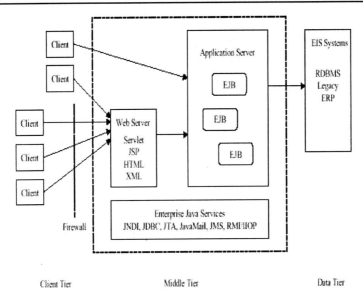

Figure 9.1: J2EE Multi-tier Architecture

With the J2EE standard, now that the different systems can speak the same language, the method they use to communicate is more and more likely to be via web protocols or Web Services (see Chapter 1). This is another positive for organizations as the primary method for communication in this J2EE world is one that almost everyone agrees on and is on many levels already accepted in the organization and, just as important, already paid for. J2EE applications and application servers are perfectly positioned to take advantage of the web services revolution.

So you might think "why wouldn't we want to use J2EE?" Well, there are several reasons why you may want to go a different direction. One may be a preponderance of legacy systems that have already been cobbled together and may be too difficult to "decouple"; the other may be Microsoft's .NET platform which is being positioned (some may say repositioned) as a direct competitor to J2EE based systems. .NET, while not exactly the same as J2EE, is Microsoft's distributed computing platform with web services baked in for good measure so it is essentially, like J2EE, a model for multi-tiered distributed computing or, in simpler terms, a way for different computing applications and platforms to speak the same language. However, J2EE is an open standard not particularly owned by one company whereas .Net is very much driven by one vendor, Microsoft.

For existing Microsoft shops –and let's face it – what company these days doesn't at least use some Microsoft software, .Net makes a lot of sense.

But it does put a lot of your marbles in one place. While enterprise software development is not yet a commodity, it is slowly moving in that direction with more and more IT development and infrastructure being outsourced to places like India and China. In this commoditized world, development tools and environment will need to get simpler and easier to use if they are to achieve mass adoption. This is one of the areas in which the J2EE specification falls down. While incredibly powerful, the specification and its development tools are still quite complex. Compare this to Microsoft's suite of Visual developer tools (the building blocks of the .NET architecture) and Microsoft wins the ease-of-use race hands down. Thus, right now, it is much easier (and cheaper) to find a train and Visual Basic developer to build applications than it is to find an accomplished Java developer. This fact alone could have serious ramifications for the mainstream adoption of the J2EE standard. Although a lot of progress is currently being made. Tools are emerging that offer a visual framework for traditional corporate application developers to use J2EE, leaving more complex programming to the elite J2EE developers.

Many of the discussions in this book revolve around the pros and cons of tying your organization to one vendor. The discussions around J2EE are no different. The current software development world is really in the process of polarizing around two development platforms, J2EE and Microsoft .NET. Neither of them are automatic winners in every case. One has to take in a lot of factors when deciding which platform to embark on. One has to ask questions like:

- What is my current dependency on one vendor and am I comfortable with that dependency?
- Am I looking for a best of breed approach or single vendor approach?
- What is the complexity of the applications I want?
- What type legacy system integration is involved?
- What type of programming resources do I have access to?
- What type of IT infrastructure investments has my organization already made?

Only by answering these questions and more can an organization really come to terms with what solution is best for their particular needs.

How will J2EE Affect Businesses?

J2EE allows organizations to take a multi-vendor approach to its application development. As long as a software vendor's product is J2EE compliant, it can be used in conjunction with other J2EE compliant suites and applications as they, in theory, all speak the same language. I say "in theory" as it is not always the practice but, in general, a J2EE compliant label means what it says. For example, a company can use a web server, application server, middleware application and database all from best of breed vendors for a particular solution using J2EE. This is not the case with, say, Microsoft .NET technology as it is difficult (but not impossible), for example, to use a J2EE Application server with Active Server Pages. Thus one is limited to using one vendor for a truly integrated system.

J2EE also provides a framework for making system integration simpler. That is, connecting, say, older legacy systems to newer online based applications. With J2EE, there is a set of tools and connectors that define how data is passed and connections are made between different systems.

Above all, J2EE, coupled with Internet technologies, truly enables a distributed computing paradigm for the enterprise. Thus geographic barriers start to erode for the computing world and J2EE starts to look like the toolkit that allows us to build the road towards OnDemand or Utility/Grid based computing (See Chapter 4).

How will J2EE Affect Consumers?

Java has moved hand in hand with the Internet in changing the way people approach their daily lives. Whether you are buying a book online, selling shares through an online broker, making a cell phone call, ordering a movie from your cable company or pulling cash out of an ATM, chances are, you're using some form of Java or J2EE.

Java has paved the way for "computing" to be embedded in almost any device independent of hardware or software from TVs to cell phones to monitoring systems in automobiles. It is likely that as the Internet and technology play more and more of a role in our lives, thus Java and J2EE will too.

What to Watch for in the Future:

No matter which way you look at it, the future for Java and J2EE is bright. In the Enterprise market, Java's ability to work with databases has made it very valuable and made it an important competitor to Visual Basic. Java can help extend the life of legacy systems. Java provides value for companies with heterogeneous computing environments due to the strength of Java as an integration tool. As computing paradigms move more and more towards a distributed, decentralized and web-based model, J2EE is perfectly positioned to be the platform of choice for a lot of this shift.

Sometime in 2004 the number of non-PC devices (PDAs, phones, TVs, watches, etc.) connected to the Internet is expected to surpass the number of PCs. This coupled with the growth in embedded systems (hardwired computer systems usually designed for a specific purpose, for example, the microprocessor controller in a car) is likely to also be a boon for J2EE and its followers as new data and connectivity applications will have to be developed.

The future of J2EE is not without its dark spots, however. J2EE is pitched in direct competition with Microsoft .NET technology and as history has shown us, competition with Microsoft is not something to be taken lightly. J2EE also faces complexity issues, many people feel. The development tools and sheer weight of the specification preclude many developers and organizations from using it effectively. J2EE's continued success and development is also tied to a consortium of different companies and bodies and not under direct supervision of one company like the .NET architecture. This means their continued progress depends on the cooperation and agreement of the involved parties, which usually means a slower time to market for updates and new features and a temptation for individual vendors to go outside the specification for a quicker time to market. Even with some of these pitfalls, Java and J2EE are likely to be one of the main pillars of this continued, connected digital revolution we find ourselves in.

Most Frequently Asked Questions

What's the difference between Java and J2EE?

J2ee is basically the enterprise, multi-tiered, distributed version of the Java programming language.

Is J2EE a technology?

Microsoft .Net

Strictly, J2EE is a set of standards and protocols that provides the functionality for developing multi-tiered applications.

What is the difference between J2EE and Microsoft .NET?

J2EE is a set of tools and standards based around the Java language for distributed, multi-tiered enterprise applications steered by a coalition of technology companies and standard bodies. .NET is essentially an initiative by Microsoft to provide a way of connecting all its tiers of software and suites into one vertically integrated, distributed package for development of the same type of enterprise applications as the J2EE standard.

Does one company own J2EE?

No. J2EE is an open standard but it was developed by Sun Microsystems, the license holder for Java, who along with other members of the J2EE consortium, continue to guide development of the standard.

Buzz Words and Definitions:

Embedded System- A combination of computer hardware and software, and perhaps additional mechanical or other parts, designed to perform a dedicated function. In some cases, embedded systems are part of a larger system or product, as in the case of an antilock braking system in a car.

Java- A high-level programming language developed by Sun Microsystems in 1995. Compiled Java code can run on most computers because Java interpreters and runtime environments, known as *Java Virtual Machines* (VMs), exist for most operating systems, including UNIX, the Macintosh OS, and Windows.

Applet - A component that typically executes in a web browser, but can execute in a variety of other applications or devices that support the applet-programming model.

Java 2 Platform, Enterprise Edition (J2EE) - An environment for developing and deploying enterprise applications. The J2EE platform consists of a set of services, application programming interfaces (APIs), and protocols that provide the functionality for developing multi-tiered, web-based applications.

Java 2 Platform, Micro Edition (J2SE) - A highly optimized Java runtime environment targeting a wide range of consumer products, including pagers, cellular phones, screen phones, digital set-top boxes and car navigation systems.

Java 2 Platform, Standard Edition (J2SE) - The core Java technology platform.

Java 2 SDK, Enterprise Edition (J2EE SDK) - Sun's implementation of the J2EE platform. This implementation provides an operational definition of the J2EE platform.

Java Message Service (JMS) - An API for using enterprise-messaging systems such as IBM MQ Series, TIBCO Rendezvous, and so on.

Java Naming and Directory Interface (JNDI) - An API that provides naming and directory functionality.

Java Transaction API (JTA) - An API that allows applications and J2EE servers to access transactions.

Java Transaction Service (JTS) - Specifies the implementation of a transaction manager, which supports JTA and implements the Java mapping of the OMG Object Transaction Service (OTS) 1.1 specification at the level below the API.

JavaBeans Component - A Java class that can be manipulated in a visual builder tool and composed into applications. A JavaBeans component must adhere to certain property and event interface conventions.

Java Server Pages (JSP™) - An extensible web technology that uses template data, custom elements, scripting languages, and server-side Java objects to return dynamic content to a client. Typically the template data is HTML or XML elements, and in many cases the client is a web browser.

JDBC™ - An API for database-independent connectivity between the J2EE platform and a wide range of data sources.

Enterprise Java Beans (EJB™) - Component architecture for the development and deployment of object-oriented, distributed, enterprise-level applications. Applications written using the Enterprise JavaBeans architecture are scalable, transactional, and secure.

CORBA - Common Object Request Broker Architecture. A language independent, distributed object model specified by the Object Management Group.

Microsoft .Net

CSS - Cascading Style Sheet. A style sheet used with HTML and XML documents to add a style to all elements marked with a particular tag, for the direction of browsers or other presentation mechanisms

Distributed Application - An application made up of distinct components running in separate runtime environments, usually on different platforms connected via a network. Typical distributed applications are two-tier (client-server), three-tier (client-middleware-server), and multi-tier (client-multiple middleware-multiple servers).

Enterprise Information System - The applications that comprise an enterprise's existing system for handling company-wide information. These applications provide an information infrastructure for an enterprise. An enterprise information system offers a well defined set of services to its clients. These services are exposed to clients as local and/or remote interfaces. Examples of enterprise information systems include: enterprise resource planning systems, mainframe transaction processing systems, and legacy database systems.

Component - An application-level software unit supported by a container. Components are configurable at deployment time. The J2EE platform defines four types of components: enterprise beans, web components, applets, and application clients.

Servlet - A Java program that extends the functionality of a web server, generating dynamic content and interacting with web clients using a request-response paradigm.

The Bottom Line:

The bottom line is that J2EE is a tremendously powerful framework for creating enterprise strength applications for the networked world. Although not without its weaknesses, it will continue to evolve and it will likely play a major role in how new applications are built and how computers are connected to each other and the Internet for a long time to come.

Top Ten Cheat Sheet: J2EE

● Java began life in the early Nineties as a platform-independent programming language created by a small group of engineers at Sun Microsystems.

● Java has become a de facto standard for Internet and enterprise computing.

● The Java 2 Enterprise Edition (J2EE) is a set of standards based on porting the Java programming language to multi-tiered enterprise applications.

● At its simplest, J2EE is an attempt to provide disparate software systems on different platforms with a common method of communication.

● J2EE allows organizations to pursue a best of breed approach when choosing software vendors.

● J2EE competes directly against Microsoft .NET technology for the enterprise application market.

● The development and evolution of Java and J2EE is guided by a coalition of interested parties led by Sun Microsystems and other tech heavyweights such as IBM and BEA Systems.

● J2EE, while widely regarded as very powerful, is also widely regarded as very complex and requires a much higher level of skills to develop than say the .NET framework.

● J2EE should benefit from the shift to distributed and Internet based computing.

● J2EE is not right for every organization. There are many factors hat should be considered before adoption, from current IT infrastructure investment to access to the right development resources.

10

Microsoft .Net

"In theory, there is no difference between theory and practice. But, in practice, there is."

- Jan L.A. van de Snepscheut

So what is Microsoft .NET (pronounced dot net)?

According to Microsoft: ".NET is the Microsoft solution for web services, the next generation of software that connects our world of information, devices, and people in a unified, personalized way."

It is important to note that unlike, say, J2EE, .NET is not a set of standards but rather a collection or interconnection of Microsoft platforms and technology to leverage the web services direction that software seems to be heading in.

While .NET and J2EE may have had rather different beginnings, they have both been commandeered to position themselves as web services development platforms. It is hard to find a statement about .NET and J2EE without the ubiquitous mention of web services (see Chapter 1).

Detailed Summary:

.NET connects Microsoft's existing product suite and repositions it as an end-to-end solution under the unified banner of web services.

Think of the .NET strategy as the glue that connects or will connect their product suite together using Internet technologies.

In that respect, Microsoft .NET can be viewed as just as much, if not more so, a product development strategy than a great technological advancement.

Some cynics would say that it is only a marketing strategy and all the technologies have previously existed.

The .NET strategy can be broken into 4 key components

- Smart Clients (PCs, cell phones, PDAs, etc.)
- Web Services
- Servers
- Developer Tools

with the .NET framework being the glue that integrates all these pieces together (see diagram below).

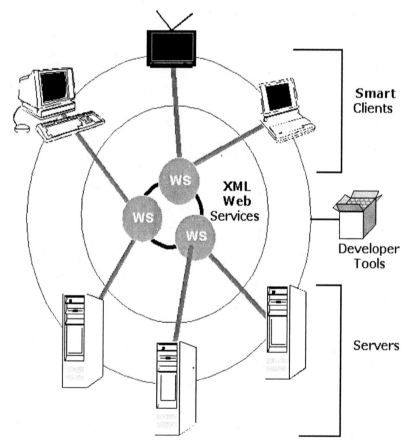

Figure 10.1 High Level .NET Overview

.NET Microsoft is aiming to leverage their existing products (many of which are already embedded in a large amount of organizations) to own all the pieces in the web services pie (and if you believe that web services are the future of enterprise application development then the strategy leads to owning the entire enterprise application development pie).

Let's take a look at the components that make up the server component of the .NET strategy, also known as the windows server system.

Windows Server System

The Microsoft Windows Server System integrated server software provides infrastructure for building, deploying, and operating Web services. They include support for XML and business-process orchestration across applications and services. Key Windows Server System products include:

- **Microsoft Application Center 2000** to deploy and manage Web applications.

- **Microsoft BizTalk® Server 2002** to build XML-based business processes across applications and organizations.

- **Microsoft Commerce Server 2002** for quickly building e-commerce solutions.

- **Microsoft Content Management Server 2002** to manage content for dynamic e-business websites.

- **Microsoft Exchange Server 2000** to enable messaging and collaboration anytime, anywhere.

- **Microsoft Host Integration Server 2000** for bridging to data and applications on mainframe legacy systems.

- **Microsoft Internet Security and Acceleration Server 2000 (ISA Server)** for SSL-secured, fast Internet connectivity.

- **Microsoft Mobile Information Server 2002** to enable application support by handheld devices, such as mobile phones.

- **Microsoft Operations Manager 2000** to deliver enterprise-class solutions for operations management.

- **Microsoft Project Server 2002** provides an extensible technology platform to develop and deploy best practices for project management across an organization.

- **Microsoft SharePoint™ Portal Server 2001** to find, share, and publish business information.

- **Microsoft SQL Server™ 2000** to store, retrieve, and analyze structured XML data.

As you can see, this is an impressive list of technologies and, on paper at least, Microsoft has as close to a full solution for Web Services and online application development as anyone. And this is just the server tier. On the client front, Microsoft has the market leader Windows, with its latest incarnation XP, completely .NET enabled. Microsoft also has Windows CE, which has been rebranded Windows CE .NET, which it is pushing aggressively for everything from mobile devices to TV set top boxes.

An important element of the "Smart Client" strategy is what's called the **.NET Framework**. This is essentially a component added to the Windows environments to make it easier for developers to build applications that connect with Web Services and integrate with other networked systems. Think of the .NET Framework as "network enabling" the operating systems and development tools. An example would be, say, you have written an Anti-Virus program for Windows and you want that program to be able to connect to the web and look for updates periodically. Well, that program would most likely use a web service of some sort to check for updates. The .NET Framework would allow you to connect to that Web Service using pre-existing code libraries.

The .NET Framework consists of two main elements: the common language runtime and the .NET Framework class library. These can be broken down as follows:

Common Language Runtime. Provides the common services for .NET Framework applications. Programs can be written for the common language runtime in just about every language, including C, C++, C#, and Microsoft Visual Basic®, as well as some older languages such as Fortran. The runtime simplifies programming by assisting with many mundane tasks of writing code. These tasks include memory management – which can be a big generator of bugs – security management, and error handling.

.NET Framework class library. The library includes prepackaged sets of functionality that developers can use to more rapidly extend the capabilities of their own software. The library includes three key components:

- ASP.NET to help build Web applications and Web services.
- Windows Forms to facilitate smart client user interface development.
- ADO.NET to help connect applications to databases.

Also, with its Visual Studio suite of developer tools, Microsoft has probably the most sophisticated and widely used development environment around. In the chapter on J2EE (Chapter 9), I discuss how it is seen by many as a weakness for the Java camp not having such high-level development tools.

Rounding out the .NET picture is the suite of Web Services that Microsoft is beginning to provide, everything from MSN to Passport to Mappoint and bCentral. By being one large organization and directly owning so many different pieces of the puzzle, Microsoft can have a distinct advantage over the competition. For the J2EE camp to create such an end-to-end solution, many different vendors will have to get together and agree on many different things.

Conversely, because it has such access to a wide array of different technologies and players, the J2EE camp can arguably field a stronger, more robust and flexible product at the end.

All in all, while many companies will be forced to choose sides, many more, in the long run, will use a combination of these technologies and that is the beauty of web services, as long as my application of software can speak XML and we can all agree on a few common standards then it doesn't really matter what platform I'm running on. It does matter, however, what kind of IT personnel I have access to and what their skill sets are and that is driving the decision making process in many organizations.

But at the end of the day, the next wave may well see J2EE and .NET and whatever else is out there being all connected together under a new acronym, for example, UWS, Universal Web Services. (I just made this last one up but feel free to use it with your IT group; there's a good chance they'll nod knowingly and who knows, it may catch on!)

It would be easy to dismiss .NET as nothing more than a clever marketing ploy but this would be wrong. Firstly, not many people have become rich underestimating Microsoft and secondly, regardless of the motivations behind .NET, the interconnection of such powerful products in the Microsoft family has to lead to a powerful platform and serious completion for any competitor and should not be taken lightly

How Will .NET Affect Businesses?

In the short term, the companies most likely affected by the .NET strategy are the ones that have already invested heavily in Microsoft technologies. They are well placed to take advantage of all the synergies and benefits .NET may well bring. They will also be in a position to leverage the current IT resources if they have to support any .NET initiatives. On the other hand, if there are weaker elements to this architecture they will most likely be stuck with them as they will have locked themselves in a single vendor scenario in many cases. The single vendor approach can also create larger dependencies and lessen a company's ability to negotiate.

In the longer term, .NET will probably have an affect of some sort on all organizations involved in online development. Either they will have to consider using the .NET solution or they (more likely) will have to integrate with systems and companies using this architecture. If they are in the Java camp the ramifications are likely that the increased competition will spur great improvements and additions to the J2EE standard. That would be their hope, at least!

Overall, I believe the .NET strategy will greatly benefit the business landscape as it will, on the whole, increase the momentum of the computing world towards online, distributed computing and integrated systems.

How will .NET Affect Consumers?

On the whole, while most consumers may never actually see the threads of the .NET solution, they will see many of the applications and byproducts. .NET will most likely hasten the progress toward the "connected world" or networked society that we are seeing by allowing software applications to easily connect to each other and to the web. A whole new class of "smart devices" is probably on the horizon from Coffee machines that connect to the web to reorder supplies to Fridges that tell your email to purchase your groceries to TVs that manage your phone calls and air-conditioning The list is endless and some if it is already here. Almost all electronic devices will

Microsoft .Net

eventually have a "brain" (processor) and be connected to the web. And it is highly likely that Microsoft and its .NET architecture will play a big role in all of this.

Most Frequently Asked Questions

What is .NET?

.NET is a set of integrated Microsoft technologies, which it hopes, will be used for connecting information, people, systems, and devices. This new generation of technology is based on Web services – small building-block applications that can connect to each other as well as to other, larger applications over the Internet.

Has .NET got anything to do with the Internet?

.NET has a lot do with the Internet. .NET, in its simplest sense, could be seen as a way for Microsoft to integrate all its technologies and make it easier for them to connect to the Internet.

Is .NET a technology?

Strictly speaking, .NET is a collection of pre-existing and newly developed technologies.

What's the difference between .NET and J2EE?

J2EE is a set of tools and standards based around the Java language for distributed, multi-tiered enterprise applications steered by a coalition of technology companies and standard bodies. NET is essentially an initiative by Microsoft to provide a way of connecting all of its tiers of software and suites into one vertically integrated, distributed package for development of the same type of enterprise applications as the J2EE standard.

What are "smart devices?"

A smart devices is any device such as PCs, laptops, workstations, smart phones, handheld computers, Tablet PCs, game consoles, etc., that can operate inside the "networked" universe.

Buzz Words and Definitions:

Active Server Pages (ASP)-A Microsoft technology for creating server-side, Web-based application services. ASP applications are typically written using a scripting language, such as JScipt, VBScript, or PerlScript. ASP first appeared as part of Internet Information Server 2.0 and was code-named *Denali*.

ADO (ActiveX Data Objects)-A set of COM components used to access data objects through an OLEDB provider. ADO is commonly used to manipulate data in databases, such as Microsoft SQL Server 2000, Oracle, and Microsoft Access.

ADO.NET (ActiveX Data Objects for .NET)-The set of .NET classes and data providers used to manipulate databases, such as Microsoft SQL Server 2000. ADO.NET was formerly known as *ADO+*. ADO.NET can be used by any .NET language.

.NET Enterprise Server product family-These products include Application Center, BizTalk Server, Commerce Server, Content Management Server, Exchange Server, Host Integration Server, Internet Security and Acceleration Server, SQL Server 2000, and Windows 2000 Server. Formerly known as BackOffice Server 2000.

C# (C-Sharp)-An object-oriented and type-safe programming language supported by Microsoft for use with the .NET Framework. C# (pronounced "see-sharp") was created specifically for building enterprise-scale applications using the .NET Framework. It is similar in syntax to both C++ and Java and is considered by Microsoft as the natural evolution of the C and C++ languages.

COM (Component Object Model)-A software architecture developed by Microsoft to build component-based applications. COM objects are discrete components, each with a unique identity, which expose interfaces that allow applications and other components to access their features.

Common Language Runtime (CLR)-A runtime environment that manages the execution of .NET program code, and provides services such as memory and exception management, debugging and profiling, and security.

COM+-The "next generation" of the COM and DCOM software architectures. COM+ (pronounced "COM plus") makes it easier to design and construct distributed, transactional, and component-based applications using a multi-tiered architecture.

Microsoft .Net

J# (J-Sharp)-A Microsoft-supported language for .NET. J# (pronounced "jay sharp") is Microsoft's implementation of the Java programming language. It is specifically designed to allow Java-language developers to easily transition to the .NET Framework and to create .NET applications.

.NET Framework-A programming infrastructure created by Microsoft for building, deploying, and running applications and services that use .NET technologies, such as desktop applications and Web services. The .NET Framework contains three major parts: the Common Language Runtime (CLR), the Framework Class Library, and ASP.NET.

.NET Framework Class Library (FCL)-The foundation of classes, interfaces, value types, services and providers that are used to construct .NET Framework desktop and Web-based (i.e., ASP.NET) applications.

Visual Basic .NET-A Microsoft-supported language for the .NET Framework. VB.NET is the "next generation" release of the very popular Visual Basic programming language.

The Bottom Line:

Some people say .NET is a marketing strategy by Microsoft; some people say it's a new network architecture, others say it's a way of integrating a host of Microsoft's technologies to take advantage of the Web and Web services. The answer is it is all of the above. .NET is a very powerful strategy by Microsoft that will lead ultimately to some very powerful technology solutions and may well set the tone for web service development and the move to distributed computing environments. While some of the technological components may not be up to the marketing hype, .NET is, and will become, a force to be reckoned with, particularly for the Java camp.

Top Ten Cheat Sheet: Microsoft .NET

◉ ".NET is the Microsoft solution for web services, the next generation of software that connects our world of information, devices, and people in a unified, personalized way."

- .NET integrates Microsoft's existing product suite and repositions it as an end-to-end solution under the unified banner of web services.

- .NET strategy can be broken into 4 key components: Smart Clients (PCs, cell phones, PDAs, etc.), Web Services, Servers and Developer Tools.

- A key part of the .NET solution is the .NET Framework, which can be viewed at a simple level as the "network enabling" or web service enabling the current Microsoft operating systems and development tools.

- By being one large organization and directly owning so many different pieces of the Web Service and online application development puzzle, Microsoft can have a distinct advantage over the competition.

- J2EE, the Java architecture for Web Service and online application development, has an advantage over .NET in that it is seen as promoting the best of breed approach, not locking an organization into one vendor.

- In the short term, the companies most likely affected by the .NET strategy are the ones that have already invested heavily in Microsoft technologies.

- In the longer term, .NET should spur fierce competition between the vendors that provide solutions for computing, in general resulting ultimately in better solutions for the Enterprise or possibly only one solution! Let's hope it's the former.

- Almost all electronic devices will eventually have a "brain" (processor) and be connected to the web. And it is highly likely that Microsoft's .NET architecture will play a big role in all of this.

- The .NET strategy will increase the momentum of the computing world towards online, distributed computing and integrated systems.

APPENDIX: GLOSSARY OF TERMS

.NET Enterprise Server product family- These products include Application Center, BizTalk Server, Commerce Server, Content Management Server, Exchange Server, Host Integration Server, Internet Security and Acceleration Server, SQL Server 2000 , and Windows 2000 Server. Formerly known as BackOffice Server 2000.

. NET Framework- A programming infrastructure created by Microsoft for building, deploying, and running applications and services that use .NET technologies, such as desktop applications and Web services. The .NET Framework contains three major parts: the Common Language Runtime (CLR), the Framework Class Library, and ASP.NET.

. NET Framework Class Library (FCL)- The foundation of classes, interfaces, value types, services and providers that are used to construct .NET Framework desktop and Web-based (i.e., ASP.NET) applications.

2.5G-The enhancement of GSM which includes technologies such as GPRS.

2G-The second generation of digital mobile phone technologies including GSM, CDMA IS-95 and D-AMPS IS-136.

3G-The third generation of mobile phone technologies covered by the ITU IMT-2000 family.

802.11 WLAN - A Wireless Lan specification defined by the IEEE.

802.11b - Industry standard for wireless Internet use. Operates through radio frequencies around 2.4 GHz. Common electronics like cordless phones and microwaves also operate on this frequency.

Access point- Hardware that connects to existing DSL or cable modem lines, essentially turning a wired connection into a wireless connection within a certain distance of the hardware. Access points are shared connections, so more than one person can access the Internet from them.

Active Server Pages (ASP)- A Microsoft technology for creating server-side, Web-based application services. ASP applications are typically written using a scripting language, such as JScipt, VBScript, or PerlScript. ASP first appeared as part of Internet Information Server 2.0 and was code-named *Denali*.

ADO (ActiveX Data Objects)- A set of COM components used to access data objects through an OLEDB provider. ADO is commonly used to manipulate data in databases, such as Microsoft SQL Server 2000, Oracle, and Microsoft Access.

ADO.NET (ActiveX Data Objects for .NET)— The set of .NET classes and data providers used to manipulate databases, such as Microsoft SQL Server 2000. ADO.NET was formerly known as *ADO+*. ADO.NET can be used by any .NET language.

Applet - A component that typically executes in a web browser, but can execute in a variety of other applications or devices that support the applet-programming model.

Application Service Providers-Application service providers offer an outsourcing mechanism whereby they develop, supply and manage application software and hardware for their customers, thus freeing up customers' internal IT resources.

Backdoor- A piece of software that allows for unauthorized access to a system.

Bluetooth - An open specification for wireless communication of data and voice. It is based on a low-cost short-range radio link facilitating protected ad hoc connections for stationary and mobile communication environments.

Bluetooth-A low power, short- range wireless technology designed to provide a replacement for the serial cable. Operating in the 2.4GHz ISM band, Bluetooth can connect a wide range of personal, professional and domestic devices such as laptop computers and mobile phones together wirelessly.

Bottom up-An approach to building things by combining smaller components, as opposed to carving them out of larger ones (top down), as is done in current photolithographic approaches to making silicon chips.

Business Intelligence-Business intelligence refers to the type of granular information that line managers seek as they analyze sales trends, customer buying habits and other key performance metrics of an organization.

Business Metrics-Business Metrics is a set of traditional and nontraditional business measurements - such as judging product and service quality, rating customer relationships and measuring employee satisfaction and commitment - that are seen as critical for improving a company's bottom line.

C# (C-Sharp)- An object-oriented and type-safe programming language supported by Microsoft for use with the .NET Framework. C# (pronounced "see-sharp") was created specifically for building enterprise-scale applications using the .NET Framework. It is similar in syntax to both C++ and Java and is considered by Microsoft as the natural evolution of the C and C++ languages.

Cell-The area covered by a cellular base station.

COM (Component Object Model)- A software architecture developed by Microsoft to build component-based applications. COM objects are discrete components, each with a unique identity, which expose interfaces that allow applications and other components to access their features.

COM+- The "next generation" of the COM and DCOM software architectures. COM+ (pronounced "COM plus") makes it easier to design and construct distributed, transactional, and component-based applications using a multi-tiered architecture.

Command Line -A space provided directly on the screen where users type specific commands. In Linux, you open a shell prompt and type commands at the command line, which generally displays a $ prompt at the end

Common Language Runtime (CLR)- A runtime environment that manages the execution of .NET program code, and provides services such as memory and exception management, debugging and profiling, and security.

Component - An application-level software unit supported by a container. Components are configurable at deployment time. The J2EE platform defines four types of components: enterprise beans, web components, applets, and application clients.

Computer Telephony Integration (CTI)-Computer Telephony Integration (CTI) is an architecture that allows telephone applications and services to be merged with computer applications and offered on open, accessible platforms. CTI simply coordinates the actions of telecommunications and computer systems to link the real-time call control capability of the telephone with the stored intelligence of computers.

Convergent assembly-Small parts to obtain larger parts, then fastening those to make still larger parts, and so on.

CORBA- Common Object Request Broker Architecture. A language independent, distributed object model specified by the Object Management Group.

CRM Applications- CRM applications can include capabilities such as sales force automation, employee help desk, customer support, and relationship enhancement. Usually, CRM applications implement CTI capabilities, giving them the ability to "pop" the appropriate screen, based on the customer's phone number (in the case of a telephone call) or the information collected from an IVR.

CSS - Cascading Style Sheet. A style sheet used with HTML and XML documents to add a style to all elements marked with a particular tag, for the direction of browsers or other presentation mechanisms.

Customer Acquisition Costs- Customer acquisition costs are the marketing and advertising expenses needed to turn a prospective customer into an actual customer. This can include money spent to entice a visitor to a website or to a bricks and mortar store to purchase goods or services.

Customer Retention- The means of satisfying a customer through a high quality level delivery of service and retaining their loyalty to purchase future products and services.

Diamondoid-Structures that resemble diamond in a broad sense, strong stiff structures containing dense, three dimensional networks of covalent bonds; diamondoid materials could be as much as 100 to 250 times as strong as titanium, and far lighter.

Distributed Application - An application made up of distinct components running in separate runtime environments, usually on different platforms connected via a network.

Typical distributed applications are two-tier (client-server), three-tier (client-middleware-server), and multi-tier (client-multiple middleware-multiple servers).

ebXML (electronic business Extensible Markup Language)-A modular suite of specifications for standardizing XML globally in order to facilitate trade between organizations regardless of size. The specification gives businesses a standard method to exchange XML-based business messages, conduct trading relationships, communicate data in common terms and define and register business processes.

eCRM- Is the acronym for electronic Customer Relationship Management. This is the online version of Customer Relationship Management utilizing and interfacing business processes and data with offline, back end systems.

EDGE-Enhanced Data rates for GSM Evolution; EDGE uses a new modulation schema to enable theoretical data speeds of up to 384kbit/s within the existing GSM spectrum. An alternative upgrade path towards 3G services for operators, such as those in the USA, without access to new spectrum. Also known as Enhanced GPRS (E-GPRS).

Embedded System- A combination of computer hardware and software, and perhaps additional mechanical or other parts, designed to perform a dedicated function. In some cases, embedded systems are part of a larger system or product, as in the case of an antilock braking system in a car.

Enterprise Information System - The applications that comprise an enterprise's existing system for handling company-wide information. These applications provide an information infrastructure for an enterprise. An enterprise information system offers a well defined set of services to its clients. These services are exposed to clients as local and/or remote interfaces. Examples of enterprise information systems include: enterprise resource planning systems, mainframe transaction processing systems, and legacy database systems.

Enterprise Java Beans (EJB™)- Component architecture for the development and deployment of object-oriented, distributed, enterprise-level applications. Applications written using the Enterprise JavaBeans architecture are scalable, transactional, and secure.

Fabricator-A small nano-robotic device that can use supplied chemicals to manufacture nanoscale products under external control.

Firewall-A firewall is a set of related programs, located at a network gateway server, that protects the resources of a private network from users from other networks. Basically, a firewall, working closely with a router program, filters all network packets to determine whether to forward them toward their destination.

Floating Point Operations- This is a method of encoding real numbers within the limits of the finite precision available on computers. Using floating-point encoding, extremely long numbers can be handled relatively easily.

GHz- A unit of frequency equal to one billion Hertz per second

Globus Alliance- A collaborative academic project centered at Argonne National Laboratory focused on enabling the application of grid concepts to computing.

Globus Toolkit- The Globus Toolkit is an open source software toolkit used for building grids. The Globus Alliance and many others all over the world are developing it. A growing number of projects and companies are using the Globus Toolkit to unlock the potential of grids for their cause

GNOME (GNU Object Model Environment)- A graphical desktop environment for UNIX and Linux that is designed to provide an efficient and user-oriented environment

GNU- GNU is a recursive acronym for "GNU's Not Unix.". It is a UNIX-compatible operating system using the Linux kernel developed by the Free Software Foundation. The design philosophy of GNU is to create a full-featured operating system composed of completely free software. Red Hat Linux combines several parts of GNU along with the Linux kernel.

GPRS- General Packet Radio Service; GPRS represents the first implementation of packet switching within GSM, which is a circuit switched technology. GPRS offers theoretical data speeds of up to 115kbit/s.

Grand Challenge- A problem that by virtue of its degree of difficulty and the importance of its solution, both from a technical and societal point of view, becomes a focus of interest to a specific scientific community.

Grid computing- A type of distributed computing in which a wide-ranging network connects multiple computers whose resources can then be shared by all end-users; includes what is often called "peer-to-peer" computing.

Teraflop- A teraflop is a measure of computer processing power and is a trillion (10 to the power of 12) floating-point operations per second.

GSM- Global System for Mobile communications, the second generation digital technology originally developed for Europe but which now has in excess of 71 per cent of the world market. Initially developed for operation in the 900MHz band and subsequently modified for the 850, 1800 and 1900MHz bands. GSM originally stood for Groupe Speciale Mobile, the CEPT committee which began the GSM standardisation standardization process.

IEEE - Institute of Electronic and Electrical Engineering.

Inbound & Outbound CRM- Inbound CRM covers the experience customers have when they initiate contact with a business through any channel – whether through call center, IVR or the Internet. Outbound CRM is when a business initiates contact with a customer for the purposes of maintaining its relationship with that customer.

IP Contact Center- IP stands for Internet Protocol. An IP Contact Center is a multi multi-channel contact center, designed to support customers using IP based telephony. Traditional specialized hardware such as ACD, PBX and IVR are replaced by applications on the IP Network. This results in a framework that is more flexible and provides a better platform for unified queuing and incorporating IP-based new media.

J# (J-Sharp)- A Microsoft-supported language for .NET. J# (pronounced "jay sharp") is Microsoft's implementation of the Java programming language. It specifically designed to allow Java-language developers to easily transition to the .NET Framework and to create .NET applications.

Java 2 Platform, Enterprise Edition (J2EE)- An environment for developing and deploying enterprise applications. The J2EE platform consists of a set of services, application programming interfaces (APIs), and protocols that provide the functionality for developing multi-tiered, web-based applications.

Java 2 Platform, Micro Edition (J2SE) - A highly optimized Java runtime environment targeting a wide range of consumer products, including pagers, cellular phones, screen phones, digital set-top boxes and car navigation systems.

Java 2 Platform, Standard Edition (J2SE)- The core Java technology platform.

Java 2 SDK, Enterprise Edition (J2EE SDK) -Sun's implementation of the J2EE platform. This implementation provides an operational definition of the J2EE platform.

Java- A high-level programming language developed by Sun Microsystems in 1995. Compiled Java code can run on most computers because Java interpreters and runtime environments, known as *Java Virtual Machines (VMs)*, exist for most operating systems, including UNIX, the Macintosh OS, and Windows.

Java Message Service (JMS) - An API for using enterprise-messaging systems such as IBM MQ Series, TIBCO Rendezvous, and so on.

Java Naming and Directory Interface (JNDI) - An API that provides naming and directory functionality.

Java Server Pages (JSP^{TM}) - An extensible web technology that uses template data, custom elements, scripting languages, and server-side Java objects to return dynamic content to a client. Typically the template data is HTML or XML elements, and in many cases the client is a web browser.

Java Transaction API (JTA) - An API that allows applications and J2EE servers to access transactions.

Java Transaction Service (JTS) - Specifies the implementation of a transaction manager, which supports JTA and implements the Java mapping of the OMG Object Transaction Service (OTS) 1.1 specification at the level below the API.

JavaBeans Component - A Java class that can be manipulated in a visual builder tool and composed into applications. A JavaBeans component must adhere to certain property and event interface conventions.

JDBC™ - An API for database-independent connectivity between the J2EE platform and a wide range of data sources.

KDE (K Desktop Environment)- Graphical desktop interface designed with free software tools and libraries for the free software community.

Kernel-"Kernel" refers to the core OS. It is an abstraction layer between hardware and software, which provides an environment for running software programs. It contains large chucks of C and assembly code for providing the functions of an OS. This is the part of Linux that is common for every distribution.

Linux Distribution- Linux distributions are developed based on the Linux kernel, adding enhancements, packaged with software and tools for installation and configuration. Some of the more popular distributions are Debian, Red Hat, Mandrake and SuSE.

Malware- Malware stands for Malicious software. A catch-all term for "'programs that do bad or unwanted things".'. Generally, viruses, worms and Trojans will all be classed as malware, but several other types of programs may also be included under the term.

Mass Mailer- A virus that distributes itself via email to multiple addressees at once is known as a mass mailer. Probably the first mass mailer was the CHRISTMA EXEC worm of December 1987.

Micro-Electro-Mechanical Systems (MEMS)-The integration of mechanical elements, sensors, actuators, and electronics on a common silicon substrate through microfabrication technology.

Micron-One millionth of a meter, or about 1/25,000 of an inch.

Millimeter-One thousandth of a meter, or about 1/26 of an inch.

MINIX- A small operating system that is very similar to UNIX that was written by Prof. Andrew S. Tanenbaum of Vrije Universiteit, Amsterdam, for educational purposes.

MNT-An abbreviation for molecular nanotechnology that refers to the concept of building complicated machines out of precisely designed molecules.

Nanofactory-A self-contained macroscale manufacturing system, consisting of many systems feeding a convergent assembly system.

Nanometer- One billionth of a meter; approximately the width of a single strand of DNA.

Nanoscale- A scale significantly smaller than a micron.

NNI- The US's National Nanotechnology Initiative.

Nominal Range- The range at which a systems can assure reliable operation, considering the normal variability of the environment in which it is used.

Open Source Software (OSS)- Non-proprietary software in which the software source code is available and can be adapted by users to suit their needs.

OS (Operating System)- The main control software of a computer system. The OS handles task scheduling, storage, and communication with peripherals. All applications installed on a computer system must communicate with the operating system. Linux is one example of an operating system.

Passive Tags- Passive tags contain no internal power source. They are externally powered and typically derive their power from the carrier signal radiated from the scanner.

Petabyte-This is a massive memory storage number equaling 2 to the power of 50 bytes – or 1,024 terabytes – or about 20 million four-drawer filing cabinets full of text documents!

Photolithography-The technique used to produce the silicon chips that make up modern-day computers.

RFID- Systems that read or write data to RF tags that are present in a radio frequency field projected from RF reading/writing equipment.

Roaming-A service unique to GSM which enables a subscriber to make and receive calls when outside the service area of his home network, e.g., when travelling abroad.

Router- A device that forwards data from one WLAN or wired local area network to another. The router is able to determine the fastest and most reliable way to send data from LAN to LAN.

Self-assembly-The process whereby components spontaneously organize into more complex objects.

Servlet - A Java program that extends the functionality of a web server, generating dynamic content and interacting with web clients using a request-response paradigm.

SOAP (*Simple Object Access Protocol*)-SOAP is a messaging protocol used to encode XML-based messages over the Internet. It adds and defines security and transactional information associated with the Web Services document or request

Source Code- Specially written instructions by a software programmer to create executable programs when run through a compiler or language interpreter.

Tag- The transmitter/receiver pair or transceiver plus the information storage mechanism attached to the object is referred to as the tag, transponder, electronic label, code plate and various other terms. Although transponder is technically the most accurate, the most common term and the one preferred by the Automatic Identification Manufacturers is tag.

TCP/IP (transmission control protocol on top internet protocol)- Communications protocol used to connect to a variety of different types of hosts on both private networks and carrier networks such as the Internet.

Terabyte- A terabyte is a measure of computer storage capacity and is 2 to the 40th power or approximately a thousand billion bytes (that is, a thousand gigabytes).

UDDI (*Universal Description, Discovery and Integration)-* A Web-based distributed directory that enables business to list themselves on the Internet and discover each other, similar to a traditional phone book's yellow and white pages. Basically, a directory where companies can find the service I'm looking for and list the services they provide in a Web Services format.

UNIX- Unix is a multi-user, multi-tasking network operating system developed at Bell Labs in the early 1970s. Linux is based on, and is highly compatible with, Unix. Unix is not a single operating system. It is, in fact, a general name given to dozens of operating systems by different companies, organizations, or groups of individuals. These variants of Unix are referred to as "flavors.". Although based on the same core set of Unix commands, different flavors can have their own unique commands and features, and may be designed to work with different types of hardware. Some popular "flavors" of UNIX are HP-UX, IRIX, Linux, NetBSD, OpenBSD, Solaris and Tru64.

URL (Uniform Resource Locator)- A publicly routable address for resources transmitted via the World Wide Web (WWW). URLs can be name-based (such as www.example.com) or address-based (such as 192.168.1.2).

Virus - A computer virus is a self-replicating program that explicitly copies itself and that can infect other programs by modifying them. It can also modify their environment so that a call to an infected program implies a call to a possibly evolved copy of the virus.

Visual Basic .NET- A Microsoft-supported language for the .NET Framework. VB.NET is the "next generation" release of the very popular Visual Basic programming language .

Voice Over IP (VOIP)- Voice Over IP, a term that originally described the transmission of real-time voice calls over a data network that uses IP, but currently is used to describe "anything over IP,," for example, voice, fax, video, etc

WEP- Wired equivalent privacy or wired encryption protocol, basic Wi-Fi security is used to protect data and Internet access from outside users. The encryption process uses algorithms to secure data being transferred via radio waves.

Wi-Fi- Wireless Fidelity, also known as 802.11b, a technology that uses radio waves for computers to connect to each other and to the Internet.

WLAN - Wireless Local Area Network.

Worm - Self-contained programs that break into a system via remotely exploitable security flaws. They then self-instantiate - their replication mechanism is directly responsible for their code running on new target host systems, rather than requiring some external action such as a user running a program or restarting the system as with viruses.

WSDL *(Web Services Description Language)* - An XML-formatted language used to describe a Web service's capabilities as collections of communication endpoints capable of exchanging messages. WSDL is an integral part of

UDDI, an XML-based worldwide business registry. WSDL is the language that UDDI uses. Microsoft and IBM developed WSDL jointly. In plain English, when I do find the service I'm looking for, this explains what it is and how to connect to it.

XML *(Extensible Markup Language)* - This is customizable way of creating tags to enable the definition, transmission, validation, and interpretation of data between applications and between organizations. It is essentially the format the web services document or request is in.

About the Authors

Dermot McCormack is a seasoned technologist and visionary who has been involved in launching more than 200 Internet sites since 1993 and has been heavily involved in the strategic implementation of technology for both new and established companies over the past 12 years, specializing particularly in online applications. In that period, McCormack has been involved in raising more than $53 million in venture capital.

Mr. McCormack was co-founder and chief technology officer of Flooz.com, an online gift currency. Flooz.com was the number 22 e-commerce site on the World Wide Web in December 2000. Before founding Flooz.com, Mr. McCormack was one of the early members of iVillage. He also has developed large-scale Web projects for clients that include Sony, Microsoft, and Intel. Before iVillage, he was the chief technology officer and co-founder of Inw@re Technologies, a pioneering Web company that developed Internet and intranet solutions. And before that, he worked in the software and automation industries.

In 2000, Mr. McCormack was named to Irish America Magazine's Business 100, a list celebrating achievements by Irish Americans in business.

Fergal O'Byrne is an Internet industry entrepreneur and founded Interactive Return back in 1998. This company has now grown to be one of the premier providers of online marketing services with an extensive client list in both Europe and the USA. Fergal is now Non-Executive Chairman of the company. He is a member of the Irish Internet Association, the Institution of Engineers of Ireland and a number of other international professional bodies. He has written many white papers and technical documents for top magazines and was the Internet industry commentator for a number of publications. He was short-listed for the 2003 Irish Internet Association's Net Visionary Award. He was part of the original team that helped to set up eircom.net, Ireland's largest ISP and was Content Development Manager in charge of such projects as the Doras Directory, the World's largest directory of rated websites.

The CTO/CIO Best Practices Handbook

By Mark Minevich
Former CTO of IBM Next Generation, Member of CIO Collective

The CTO/CIO Best Practices Handbooks feature need to know information at your fingertips, direct from leading industry executives. Why spend countless hours searching for relevant thought leadership articles, specific pieces of statistical data, and navigable reference information, when one resource provides it all? The CTO/CIO Best Practices Handbook offers a wealth of articles authored by leading executives, as well as vital industry statistics, essential reference material – including forms and interactive worksheets – and a list of additional field-specific resources with contact information. The book includes CTO/CIO related technology articles written by C-Level (CEO, CTO, CFO, CMO) executives from companies such as BMC, BEA, Novell, IBM, Bowstreet, Harte-Hankes, Reynolds & Reynolds, McAfee, Verisign, Peoplesoft, Boeing, GE, Perot Systems, and over 50 other companies - available exclusively from Aspatore Books. Featuring information on:

Summary of Key Leaders - Roles and Responsibilities (CTO, CIO, Chief Scientist); Fundamentals of the CIO/CTO role; Importance of the CTO/CIO Profession; Background of CTO/CIO Profession; Change and Transformation; Globalization Perspective; US Government Perspective; McKinsey Perspective; Goldman Sachs Analysis Report; Natural Maturation of Markets and Efficiency; Competitiveness Issues; Current Economic Climate and Changes; New Generation and Digital Revolution; Women as CIO/CTO's; Are CIOs in Decline; Changing Environment in Context; Outsourcing and Offshoring; Changes leading up to Mainstream Outsourcing- Offshoring model; Trends and Figures; Challenges in Offshoring; Russia; India; Offshoring Maturing; Risk Management; What does it mean for US IT market?; What does it mean for CIO; New Paradigm; New Economy- Creating Value for Customers; CTO/CIO- Change and Transformation; CTO Priorities; Top Industry Players; Emerging Technology Direction and Vision; Next Generation Consulting Report; Future Growth Opportunities and Technologies; Strategic and Influential Relationships empowering CTO - a Complex Ecosystem; CTO Strategic Roles and Responsibilities; Skills and Competencies of an Effective; CTO Technology Summary; CTO – Leadership and Coaching; CTO and the emerging and competitive world; Monitoring and Assessing New Technologies; CTO-Strategic Planning and Direction; CTO Innovation and Commercialization; CTO and Evangelist; CTO and Globalization; CTO- Merger & Acquisition; CTO – Marketing and Media role; CTO- Government, Academia, Professional & Much More...

640 Pages - $219.95

To Order or For Customized Suggestions From an Aspatore Business Editor, Please Call 1-866-Aspatore (277-2867) Or Visit www.Aspatore.com

C-Level
Technology Review
Reducing Expenses & Delivering Profits

Trying to stay a step ahead of the key issues every technology professional needs to be aware of? Interested in interacting with a community of technology experts and executives from the world's top companies? For only $1095 a year, subscribe today to *C-Level Technology Review*, the quarterly journal for the senior most intelligence with respect to the technology.

Sample C-Level executive contributors/subscribers from the following companies include:

Advanced Fibre Communications, American Express, American Standard Companies, AmeriVest Properties, A.T. Kearney, AT&T Wireless, Bank of America, Barclays, BDO Seidman, BearingPoint (Formerly KPMG Consulting), BEA Systems, Bessemer Ventures, Best Buy, BMC Software, Boeing, Booz-Allen Hamilton, Boston Capital Ventures, Burson-Marsteller, Corning, Countrywide, Cravath, Swaine & Moore, Credit Suisse First Boston, Deutsche Bank, Dewey Ballantine, Duke Energy, Ernst & Young, FedEx, Fleishman-Hilliard, Ford Motor Co., General Electric, Hogan & Hartson, IBM, Interpublic Group, Jones, Day, Reavis & Pogue Ketchum, KPMG, LandAmerica, Leo Burnett, Mack-Cali Realty Corporation, Merrill Lynch, Micron Technology, Novell, On Semiconductor, Oxford Health, PeopleSoft, Perot Systems, Prudential, Ropes & Gray, Saatchi & Saatchi, Salomon Smith Barney, Staples, TA Associates, Tellabs, The Coca-Cola Company, Unilever, Verizon, VoiceStream Wireless, Webster Financial Corporation, Weil, Gotshal & Manges, Yahoo!

C-Level Technology Review enables executives to stay one step ahead of the technology curve and participate as a member of a community of leading technology executives, as well as product and service purchasing decision makers. C-Level Technology Review is an interactive journal, the content of which is provided exclusively by its readership, and upon subscribing, new members become eligible to submit articles for possible publication in the journal. The journal has the highest concentration of C-Level (CEO, CFO, CTO, CMO, Partner) contributors/subscribers from the Global 1000 of any business journal in the world. Subscribers look to C-Level Technology Review to stay abreast of new technologies and products and services, which they can employ to increase profits, reduce costs, and streamline operations for their companies. While other technology publications focus on the past, or current events, C-Level Technology Review helps executives stay one step ahead of major technology trends that are occurring 6 to 12 months from now. Each quarterly journal is written by leading technology executives and addresses new trends, technologies, and other developments that directly impact the corporate world. Four Quarterly Issues – $1095/Year

To Order or For Customized Suggestions From an Aspatore Business Editor, Please Call 1-866-Aspatore (277-2867) Or Visit www.Aspatore.com

Technology Best Sellers

- Leading CTOs - CTOs from Peoplesoft, BMC, Novell & More on Technology as a Strategic Weapon for Your Company - $27.95

- Technology Blueprints - Strategies for Optimizing and Aligning Technology Strategy & Business - $69.95

- Software Agreements Line by Line - How to Understand & Change Software Licenses & Contracts to Fit Your Needs - $49.95

- The Software Business - CEOs from Information Builders, Bowstreet, Business Objects & more on the Business of Developing & Implementing Profitable Software Solutions - $27.95

- Profitable Customer Relationships - CEOs from Leading Software Companies on using Technology to Maxmize Acquisition, Retention & Loyalty - $27.95

Grow Your Technology Library! Buy All 5 Titles above and

Save 30% - $149.75

Buy all 5 including The CTO/CIO Best Practices Handbook and

Save 40% - $259.25

Call 1-866-Aspatore (277-2867) to Order

To Order or For Customized Suggestions From an Aspatore Business Editor, Please Call 1-866-Aspatore (277-2867) Or Visit www.Aspatore.com

Other Best Sellers

Visit Your Local Bookseller Today or
www.Aspatore.com For More Information

- The CEO's Guide to Information Availability - Why Keeping People & Information Connected is Every Leader's New Priority - $27.95

- The Ways of the Techies - CEOs From McAfee, VanDyke & More on the Future of Technology - $27.95

- Privacy Matters - Privacy Chairs & CTOs From GE, LandAmerica, McGuireWoods, Kaye Scholer & More on Privacy Strategies for Businesses - $27.95

- Software Product Management - Managing Software Development from Idea to Product to Marketing to Sales - $44.95

- The Wireless Industry - CEOs from AT&T Wireless, Arraycomm & More on the Future of Wireless Technology - $27.95

- The Semiconductor Industry - CEOs from Micron Technology, Xilinx & More on the Future of Semiconductor Technology - $27.95

- The Telecommunications Industry - CEOs from Voicestream, Primus & More on the Future of Telecommunications Technology - $27.95

- Web 2.0 - A Look at Technology in 2008 for Businesses, Consumers, Investors & Technology Professionals - $44.95

- Being There Without Going There - Managing Teams Across Time Zones, Locations and Corporate Boundaries - $24.95

- Tech Speak - A Dictionary of Technology Terms Written by Geeks for Non-Geeks - $19.95

- Kiss the Frog - Integration Projects and Transforming Your Business With the Technology of BPI - $24.95

To Order or For Customized Suggestions From an Aspatore Business Editor, Please Call 1-866-Aspatore (277-2867) Or Visit www.Aspatore.com